Solar Optical Materials

Applications & Performance of Coatings & Materials in Buildings & Solar Energy Systems

Proceedings of the Conference, Oxford, UK, 12-13 April 1988

Edited by

M. G. Hutchins

Solar Energy Materials Research Laboratory
Oxford Polytechnic, Oxford, UK

USTRATH

Published for
INTERNATIONAL SOLAR ENERGY SOCIETY, UK SECTION

by

PERGAMON PRESS
OXFORD · NEW YORK · BEIJING · FRANKFURT
SAO PAULO · SYDNEY · TOKYO · TORONTO

Other Pergamon Titles of Related Interest

ALAWI et al	Solar Energy & the Arab World
ALAWI & AYYASH	Solar Energy Prospect in the Arab World
BOWEN & YANNAS	Passive & Low Energy Ecotechniques
CARTER & DE VILLIERS	Principles of Passive Solar Building Design
ISES	Intersol 85
ISES	Advances in Solar Energy Technology
McVEIGH	Sun Power, 2nd Edition
SODHA et al	Solar Passive Building

Pergamon Journals (*free specimen copy gladly sent on request*)

Energy
Energy Conversion & Management
Heat Recovery Systems & CHP
International Journal of Hydrogen Energy
Materials Research Bulletin
Solar & Wind Technology
Solar Energy

Solar Optical Materials

Applications & Performance of Coatings & Materials
in Buildings & Solar Energy Systems

UK	Pergamon Press plc, Headington Hill Hall, Oxford OX3 0BW, England
USA	Pergamon Press Inc, Maxwell House, Fairview Park, Elmsford, New York 10523, USA
PEOPLE'S REPUBLIC OF CHINA	Pergamon Press, Room 4037, Qianmen Hotel, Beijing, People's Republic of China
FEDERAL REPUBLIC OF GERMANY	Pergamon Press, Hammerweg 6, D-6242 Kronberg, Federal Republic of Germany
BRAZIL	Pergamon Editora, Rua Eca de Queiros, 346, CEP 04011, Sao Paulo, Brazil
AUSTRALIA	Pergamon Press Australia, PO Box 544, Potts Point, NSW 2011, Australia
JAPAN	Pergamon Press, 5th Floor, Matsuoka Central Building, 1-7-1 Nishishinjuku, Shinjuku-ku, Tokyo 160, Japan
CANADA	Pergamon Press Canada, Suite 271, 253 College Street, Toronto, Ontario, Canada M5T 1R5

First edition 1988

ISBN 0-08-036613-9

C O N T E N T S

FOREWORD

Spectrally selective solar absorber surfaces are the best known example of optical materials use for enhancing the thermal performance and efficiency of solar energy systems. The use of heat mirrors to reduce thermal losses in glazings is also well known. In more recent years considerable research interest has developed in areas such as electrochromic cells for the dynamic control of window transmittance, transparent insulation, and the service lifetime of solar optical materials. New application areas such as automotive glazing and daylighting control have been found. Measurement techniques for the determination of optical properties have been refined and improved but problem areas still exist, e.g., in the spectral directional measurement of infrared properties.

The Solar Optical Materials conference provided an opportunity to assemble a number of leading European experts to discuss progress and areas of concern relevant to materials use in solar energy and buildings applications. These Proceedings contain the texts of the papers presented at the conference.

The papers are ordered within the structure of the four sessions used for the conference. The sessions are entitled:

I. Transparent media for advanced window applications.

II. Optical switching films and novel materials for the controlled concersion of solar radiation.

III. Selective absorber surfaces, durability and service lifetime prediction.

IV. Measurement techniques for surface charaterisation.

It has been a pleasure to be involved in the organisation of this conference and on behalf of UK-ISES I should like to express my thanks to Oxford Polytechnic's Conference Office and to Pergamon Press for their invaluable assistance throughout.

Dr M.G. Hutchins
Oxford Polytechnic

March 1988

PRINCIPLES AND PROPERTIES OF HEAT MIRROR COATINGS FOR DOMESTIC WINDOWS

R.P. HOWSON

PHYSICS DEPARTMENT,
LOUGHBOROUGH UNIVERSITY OF TECHNOLOGY
LOUGHBOROUGH
LEICS.   LE11   3TU

REQUIREMENTS

Transfer of energy by radiation is characterised by the temperature of the source and the receiver, and their emission and absorption characteristics. The amount of radiative energy emitted is proportional to the absolute temperature, T, to the fourth part, through Stefans constant, σ, and the emissivity, ε. $M_E = \varepsilon S \sigma T^4$. S is the area. This energy is distributed in wavelengths with a maximum given by $T\lambda_{max} = 2.9 \times 10^{-3} m^\circ K$. The amount of energy absorbed by the receiver is dependent on the incident radiant flux and the integrated energy abs rptance, A. A receiver then intercepts energy over one wavelength range, characteristic of the source temperature, and radiates it over a range characteristic of its own temperature. Because the radiation from the sun is characteristic of a source at about $5,800^\circ K$, while energy is radiated from the receiver at around room temperature, say $300^\circ K$, it is possible to adjust the optical properties of a window to act as an energy trap. The receiver window passes energy at short wavelengths through to the energy absorber, i.e., the room, with little attenuation. The visible window, which is in general opaque to heat/infra-red radiation, will radiate this heat, but characteristic of a much longer wavelength. This radiation can be limited by adjusting the surface properties to be highly reflective in that spectral region. From heat balance considerations the emissivity, ε, $= 1-R$ and it will be low and the ridiation inhibited. The heat mirror is created. It requires to have high transmittance for visible wavelengths, those of the sun's energy, and high reflectance, and hence low emissivity, for heat-infra-red wavelengths. This distribution is shown in fig. 1. A window is also to see through. The response of the human eye is also shown in fig. 1. Generally, inhibition of radiant energy loss is best done within a double or more glazed enclosure. A way of passing solar energy into an enclosure to raise its temperature and inhibiting loss of energy is thus created.

The gaining of energy for a window system is not always desirable. In some cases visual contact is required but minimum energy transfer is desired to heat an already hot room. In this case windows with good visual, but poor solar, transparency are required, and are obviously narrow band filters around the peak of the eye's sensitivity in the green region of the spectrum, at 550nm, with high reflectance elsewhere, though absorption could be tolerated with good window cooling. Solar radiation

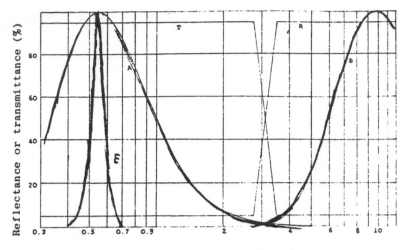

Fig. 1:   The emission spectra of the sun (5,800°K,A) and for
           surfaces at room temperature (300°K,B) normalised to
           the same peak value. Reflectances (R) and transmittances
           (T) for an ideal filter are shown together with the
           response of the eye (E).

is received over large areas and domestic windows are designed
to be as large as possible to utilise visual and heat energy.
Any coated window must be available in large areas and at a
cost which will justify their use. Materials and techniques
have recently emerged to meet these requirements, and are
commonly used in domestic windows. The background to this
development can be described to indicate where future progress
is likely to take place.

MATERIALS

Established optical filter manufacturers use absorbing dyes or
multiple layers of alternating dielectrics in an interference
array. The former does not give the required properties, the
latter can give the performance using a large number of layers,
which makes it very expensive. A filter requires to be of a
small number of layers, of little thickness, made with a technique that gives high uniformity and high rates. They need to
be surface layers which cannot be protected so that high
durability is required. The only way this performance can be
achieved is to use the intrinsic properties of a material.
It turns out that some metals have the selected properties
required and they can be enhanced in a simple way.

The theory of metals explains that their optical properties
are related to the density of free carriers and their relaxation time to give a reflection which is a function of
wavelength. It is dielectric in nature at low wavelengths,
falls to a minimum and rises to close to 1 for longer wavelengths
(fig. 2). This reflectance is associated with the highly
absorbing nature of the interaction of electrons with electro-
magnetic radiation. The transparent region at low wavelengths
is not easily seen with many materials because of the simul-

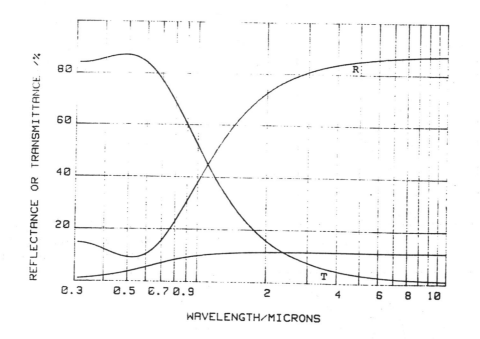

Fig. 2: The reflectance and transmittance curve for a thin
film (60nm) of an "ideal" metal.

taneous molecular type absorption of the system. Elemental
metals, which have a small molecular absorption with long
relaxation times, to have the potential for selective reflectance include silver, gold and copper. Unfortunately the
wavelength at which the transition to higher reflecting surfaces
occurs is too low, generally in the visible or UV. However,
if a high index optical matching layer is used either side of
the thin metal layer, then the transition waveband can be moved
to the infra-red and the visible transparency can be increased
to a higher value. This is done with very thin layers of each,
about 30nm for all layers. (Fig. 3).

Compound metals such as the nitrides of Ti, Zr or Hf also show
selective properties and may be used to give more durable
but less effective filters. (Fig. 4).

The metallic oxides of indium, tin, cadmium and their alloys
show the optical properties associated with less-well-conducting
metals, with a transition from transparent to reflecting occuring around 2µm, and remain free of molecular absorption in the
visible region. They show a strong dielectric reflection
associated with their high refractive index of around 2. This
can be eliminated using anti-reflection layers of intermediate
index. Of more concern is the fact that these materials have
to be of consdierably greater thickness, than the conventional
metals, to achieve similar IR reflectances and hence low
emissivity. This thickness leads to interference colours being
seen in reflection in the visible which changes with viewing
angle. This is regarded as unacceptable for domestic windows
so that material properties have either to be the best possible,
to allow films of thickness below that which reflected interference colours are seen, or thick films used, hwich give
multiple wavelength reflection peaks which are seen as "white".

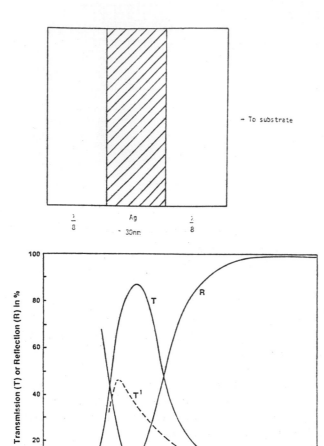

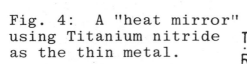

Fig. 3: The construction and performance of a three layer, oxide-silver-oxide mirror.

Fig. 4: A "heat mirror" using Titanium nitride as the thin metal.

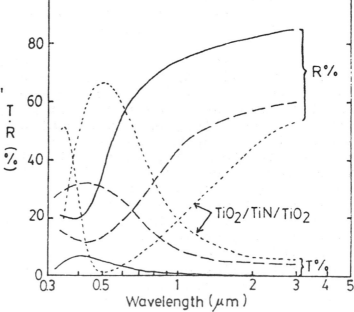

4.

Unfortunately, very thick films often have rough surfaces leading to a "milky" appearance. (Fig. 5).

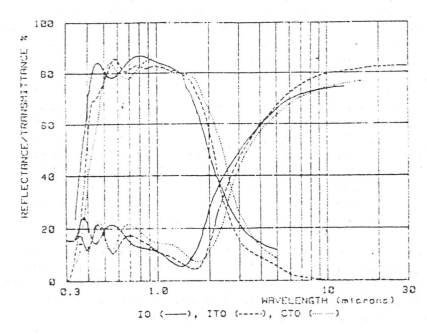

Fig. 5: The optical performance of electrically conducting oxide films.

IO (——), ITO (----), CTO (·······)

The opposite problem occurs with metal films which are required in thicknesses which are so small as to be close to the level at which they become discontinuous and lose their selective properties. The technique of providing them can be an important influence on their properties. The trade in properties that is done is shown in fig. 6, calculated for "bulk" properties and also those actually realised.

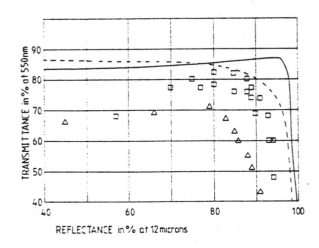

Fig. 6: The theoretical performance of simple silver films (---) and 3-layer sandwiches (——) compared with actual performances (Δ ☐ respectively)

TECHNIQUES

Oxide coatings are traditionally made by the pyrolytic decomposition of a vapour, solution in a spray, of a metal compound in the presence of oxygen. Doping to produce the required properties can be done with the addition of the dopant in a similar way. Typical films made in this way are $Sn(:Fl)O_2$ and $In(:Sn)_2O_3$. These processes require high temperatures of around 400°C and problems are encountered with the diffusion of sodium ions from soda-lime glass into the films, which destroy their electrical conducting characteristics and hence their IR reflectivity. Barrier layers have to be used. Within the last few years, the technique of planar magnetron sputtering has emerged to create a method of producing both thin metal and metal oxide films with the desired properties onto large area substrates at a higher rate. This technique is used at the moment to produce oxide-Ag-oxide filters which form the basis of the current commercial market and is being used for indium-tin oxides as well. Planar magnetron sputtering is a vacuum technique which is very inefficient in its use of energy. It requires expensive capital equipment and some skill to control. It can, however, provide a large area film at sufficient rate to make the financial and energy costs small, making the coating an essential feature of any window being replaced for some other reason. The add-on cost is justified.

Sputtering is a process which provides good adhesion of coatings to unheated substrates and can be used to coat both rigid and flexible polymers with coatings to give desirable electrical and optical properties, but also to provide abrasion resistance which may give greater impetus to their wider use.

EXTENSIONS

Double glazing incorporating heat mirrors is now commonplace, but this is only the beginning. Many climates and applications require solar energy blocking whilst maintaining visual trans-parency; after all, the highest energy consumption in many parts of the USA occurs in summer due to the use of air conditioning. Such a filter uses a combination of Fabry-Perot techniques with the intrinsic properties of a thin metal layer which can be realised in a 5-layer coating of oxide-metal-oxide-metal-oxide (fig. 7). The centre layer provides the Fabry-Perot transmission selection of whichever order and wavelength is chosen; the metal provides the IR reflection.

More and more features are required of windows of glass, E-M radiation rejection, ability to be tempered, heating of the surfaces to prevent condensation etc. These properties need to be extended to polymers.

CONCLUSIONS

Current window coatings for heat mirrors generally consist of SnO_2-Ag-Al-SnO_2. The tin oxide coatings are about 400Å thick and are produced by reactive planar magnetron sputtering; silver and aluminium are directly sputtered, the aluminium being made very thin, and it is there to protect the silver during the final deposition process for the oxide. Such coatings show very little colour in reflection, good transmittance and an IR reflectance of up to 90%, i.e. an emissivity of 0.1. These properties are all that are needed essentially to eliminate the radiation component part of the heat conduction of a double glazed window when it is placed on one internal surface. The U value is dominated by convection and edge loss through the sealant. These coatings are, however, soft and sensitive

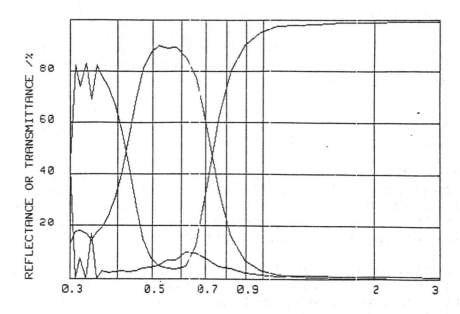

Fig. 7:   The predicted performance of a 5-layer filter.

to water vapour and can only be used in sealed double glazing
systems where they are kept in a dry atmosphere and cannot
be touched.   More durable and environment resistive coatings
with similar properties are required for double glazing.
Higher optical performance coatings will find application in
solar heat collectors.   Perhaps the biggest requirement is
for a low emissivity coating for a window which provides visual
transparency but little solar energy input, preventing the
glare and extreme temperatures encountered in rooms subject
to direct solar radiation, but holding the heat or cold
inside which is provided seperately through solar conversion
or environmental heating/cooling.

GENERAL REFERENCES

1. Fan, J.C.C. and Bachner, F.J., Transparent heat mirrors
    for solar-energy applications, Applied Optics Vol. 15,
    pp. 1012-1017 (1976)

2. Proc. SPIE, Vol. 324, Optical Coatings for Energy
    Efficiency and Solar Applications, Jan. 28-29th, 1982,
    Los Angeles, Ca., USA

3. Proc. SPIE, Vol. 438, Optical Materials and Process
    Technology for Energy Efficiency and Solar Applications,
    Aug. 23-25th, 1983, San Diego, Ca., USA, and later
    in this series. The proceedings are also published in
    Solar Energy Materials, North-Holland - Amsterdam.

4. C.A. Bishop and R.P. Howson, The performance of large
    area optical filters using DC magnetron sputtered metal
    thin films in oxide-metal-oxide sandwiches, Solar Energy
    Materials, 13, 10 (1986)

INDUSTRIAL PRODUCTION OF LOW EMISSIVITY COATINGS

M.J. Gainsbury.

Everest Double Glazing
Lodge Farm Industrial Estate
Harlestone Road
Northampton

INTRODUCTION

For over 3 years Everest Double Glazing have sputtered low emissivity coatings onto architectural float glass.

This paper describes the problems of larger scale production of sputtered thin films.

ARCHITECTURAL GLASS COATINGS.

There are dozens of architectural glass coating systems in use throughout the world.

The coatings currently of interest to the domestic U.K. and North European glazing markets are mostly low emissivity. There are two principle methods of industrial production of lowemissivity coatings onto float glass:

i.   Pyrolytic. In this process the coating is sprayed onto hot glass normally during the glass making process. It potentially has a very high production rate compared with sputtered thin films but its use is normally restricted to glass producers; currently in Europe both Glaverbel (Belgium) and St. Gobian (France) are producing low-E coatings by this method and marketing them in Northern Europe. Emissivities achieved by this method are in the order of 0.19 to 0.4 and the lower emissivity value glass tends, in Everest's experience not to offer high enough optical clarity in transmission for use in the U.K. domestic glazing market.

ii. Sputtered Thin Films. A thin film is sputtered onto the glass surface using dc magnetrons. Typical film thickness is in the order of 100nm. This process can produce coatings having emissivities in the order of 0.14 and still produce transmission qualities in the visible light range that are acceptable to the domestic market. The process is slower than the Pyrolytic but it is possible to produce a complete family of coatings in the one plant and you do not need to own a float plant !

LOW E COATING TYPES.

A number of different materials have been used to produce architectural low-E coatings. The earliest commercial coatings were based on a gold film. Pilkingtons, for a number of years, favoured copper based coatings and their original kappafloat was copper based. Most sputtered thin film coatings in Northern Europe are currently silver based. The coating in use by Everest is a leybold AG system based on a 10nm silver coating giving an emissivity

value of 0.14.   The coating system is based on 4 layers as shown
in Figure 1.

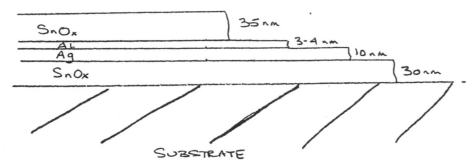

Fig. 1.   SILVER BASED LOW-E COATING (LEYBOLD AG)

The Aluminium layer is included as a diffusion barrier to prevent loss of
the silver layer   during application of the final tin oxide layer.   This
particular layer system gives a good emissivity value and only a small
reduction in transmission,   transmission is typically 84% at 550 nm.   In
reflection the colour can vary from red to blue depending on the relative
thicknessess of the tin oxide layers.   Everest standard is 'blue side of
neutral' as it is with other users of this film system.

PRODUCTION OF LOW E GLASS.

Outline description of Coating Machine.

The Everest Production Coater in common with most modern glass coaters
is a horizontal load lock machine having a 3210 x 2500 mm load size.   The
substrate is batched through a series of load lock chambers into a continuous
process chamber.   The machine described has a entrance chamber, glow
discharge chamber, 6 cathodes and an exit chamber.   The whole machine has
a volumetric capacity in exess of 35 cubic metres and is some 35metres long.
Process capacity is over 750,000 square metres of float per annum on 3 shift
continuous operation for 5 days a week.

The machine pumping scheme is based on cascade roots pumps backed by
rotary vane pumps.   High vacuum pumping is by oil-diffusion and turbo-
molecular pumps backed by the roots pumps.   All the diffusion pumps are
fitted with cold caps and water cooled baffles to minimise back streaming.
The process chamber is nitrogen vented for maintenance access and all
chambers are heated to speed pump down to process pressures.   Typically
pumpdown and outgassing takes 6-10 hours, process start pressure being
10^{-3} mbar.

Preparation of the Substrate.

For the successful production of Low-E glass the substrate must be high quality and in a fresh condition. The common practice is to specify mirror quality glass and under 28 days old. The glass is prewashed in a hot detergent wash to remove glass cutting oil residues, interleaving lucite and any other dirt that the glass surface has been contaminated with. After the prewash, glass is scrubbed with a cerium oxide slurry to polish the surface. The final wash is a 3 stage de-ionised water process the final rinse water conductivity being in the order of 5 micro Siemens. After washing, the glass is dried through air knives, static discharge and onto the heated inlet conveyor to await transport into the coater entrance chamber.

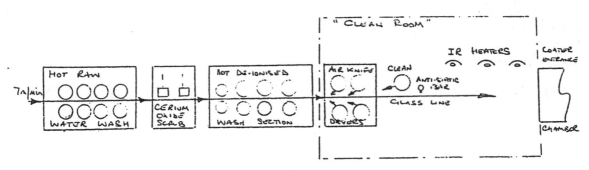

Fig. 2.   SCHEMATIC -SECTION THROUGH GLASS WASHER

The whole washing process takes 2 minutes for a 3 metre long sheet of glass.

After washing, each sheet of glass or load of pieces, (3180x2500) waits on the entrance chamber buffer under the infra-red lamps. When the entrance chamber is free and at atmospheric pressure the entrance valve is opened and the glass transported at 30m/min. into the chamber. The valve is shut and wedged and the chamber pumped down to 10^{-3} mbar. At this pressure the glass is tansfered into the second chamber. Again the inter-connecting valves are shut, Argon gas admitted the pressure raised to 10^{-2} mbar and 1500 volts applied to the glow discharge cathodes. This glow discharge is continued for 90 seconds before the chamber is pumped down to 10^{-4} mbar ready for transfer into the sputter chambers. see Fig.3

Sputtering Chambers.

Once into the main process chamber the glass is transported at constant speed (typically 1m/min.), each load being stacked about 150mm apart nose to tail. The initial tin oxide film is applied by 2 cathodes each as shown in fig. 4.

11.

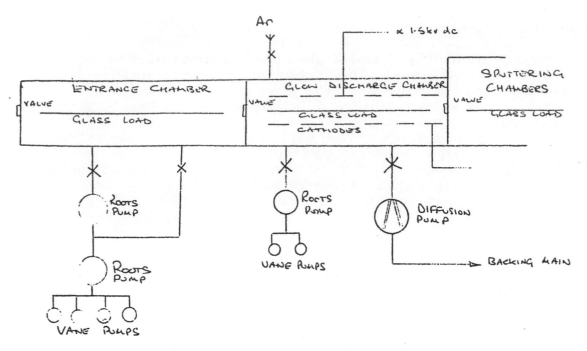

Fig. 3. GLASS PREPARATION - ENTRANCE AND GLOW DISCHARGE

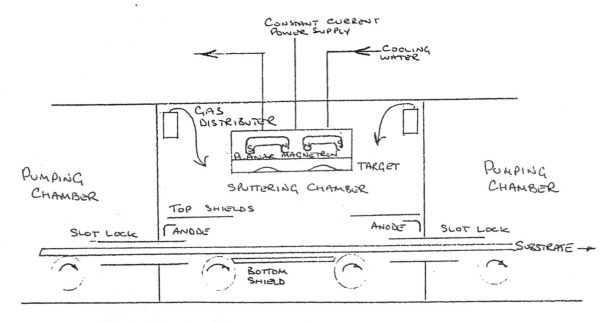

Fig 4. GENERAL LAYOUT OF A SPUTTERING CATHODE

Typically at 1m/min. and a shielding gap of 220mm, 30nm of tin oxide is sputtered with 2 cathodes each running at 35 amps with an Argon/Oxygen flow of 240ml/min. Typical cathode voltages are 400-450 volts (dc) and chamber pressure 2-5 x 10^{-3} mbar.

Silver and aluminium are sputtered in pure argon, typical flow rates being 40ml/min. To reduce the quantity of oxygen in the non-reactive chambers they are seperated from the tin oxide chambers by long dynamic slots, two diffusion pumped chambers and a turbo-molecular pump chamber. Each layer is applied sequentially as shown in fig.5.

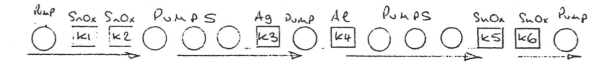

Fig. 5. SCHEMATIC - CATHODE LAYOUT.

Exit from Process Chamber.

After the tailend of the glass load has cleared the final tin oxide cathode (K6) the load is accelerated into the exit chamber. The exit chamber is then sealed vented to atmosphere and the load discharged onto the inspection conveyors. The exit chamber is then sealed and pumped down to 10^{-4} mbar ready for the next load to emerge from K6. This pump down has to be completed within 2 minutes in order to be ready in time. Failure of the exit system to be ready causes the machine transport system to break up with the loss of all the machine contents.

Determination of Coating Characteristics.

The coating optical characteristics are measured using a Pacific Scientific Spectro-gard colour measurement system. All measurements are reflective spectral inclusive. The CieLab system is used, the sample characteristics being compared with the plant standard data.

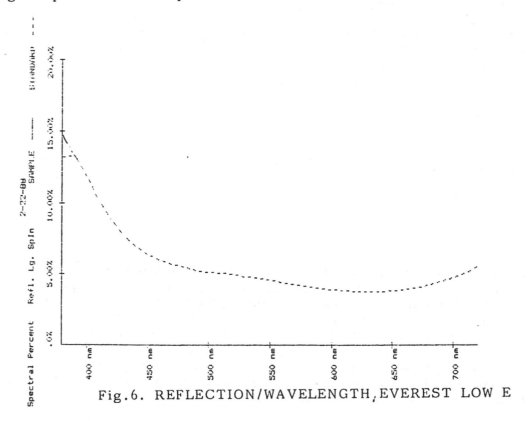

Fig.6. REFLECTION/WAVELENGTH, EVEREST LOW E

The sheet resistance of the silver layer is determined using a 4 point probe. Typical sheet resistance for a 0.14 emissivity film is 10-12 ohms/square Direct emissivity measurement is made using a direct reading emissometer. Using the data obtained from these tests adjustments are made to the machine operating parameters.

PROCESS PROBLEMS.

Stretching the sputtering process to the scale of a production machine produces some interesting process problems.

Stability.

With relatively large chambers with frequently opened lock valves and different configurations of substrate load there is a regular variation in cathode pressures. Each pressure fluctuation has a resultant variation in cathode voltage. These fluctuations together with oxide deposition on the target surface eventually trigger cathode arcing. The arcing, particularly with the oxide sputtering cathodes will eventually lower the deposition rate so much that the process has to be abandoned and the targets cleaned or changed. Since the mass-flow controllers currently on the machine are comparatively slow the reactive sputtering chambers are supplied with gas at a specified flow-rate "flooding" the chambers with oxygen to ensure that there is sufficient oxygen available for the growing film. Such practice leads to excess oxygen levels in the machine creating problems in the aluminium and silver cathodes.

Film Consistancy.

Some variation of film properties are seen across the width of the machine. Major variations are caused by:

i. Inconsistancy of crystal structure of silver targets. Obtaining a reliable supply of silver targets with a consistant crystal size across the length and depth of the target plate is difficult. (Target size is 3400x200x18 thick) variations of crystal size result in a differential sputtering rate across the width of the coaters.

ii. Gas distribution. The cathode gas supply is metered by the mass flow controllers and fed into 2 gas distribution pipes per chamber. The distribution pipes have 1.5mm diameter holes spaced evenly along the length of the pipes. This arrangement does not give an effective even distribution across the chamber width resulting again in differential sputtering yeild.

THE FUTURE.

Given better control systems such as plasma emission monitoring and high speed mass flow controllers, better control of the reactive sputter gas pressures can be achieved. Such control should enable more stable and longer target operation to be obtained and open up the possibilties of producing optical quality Low-E coatings from such films as TiN etc.

Improvements to the pumping schemes by the inclusion of cryogenic pumping would further improve the pump down time and reduce the start-up losses due to high water vapour content of the chambers.

PREPARATION AND PROPERTIES OF ZINC OXIDE FILMS FORMED BY RF SPUTTERING

C. C. Figgures and M. G. Hutchins
Solar Energy Materials Research Laboratory
School of Engineering
Oxford Polytechnic
Oxford
England

ABSTRACT

RF sputtering with substrate bias of zinc oxide thin films for heat mirror applications has been performed. Both zinc and zinc oxide targets have been used with Ar/O_2 and Ar/H_2 reactive gases respectively. Film properties have been characterised by optical, electrical and microstructural studies. Transparent zinc oxide films have been produced which are too electrically resistive for heat mirror use. Black zinc oxide films have also been made which show favourable properties as selective absorbers.

INTRODUCTION

Zinc oxide is a II-VI n-type defect semiconductor which can be made electrically conducting by creating deviations from stoichiometry via zinc interstitials and oxygen vacancies. It has a wide enough bandgap (3.2 eV) to make it transparent to most of the solar spectrum. Compared to presently used indium and/or tin based coatings zinc oxide is low cost and relatively abundant as well as being non-toxic. This makes it a candidate material for heat mirror applications. Several papers on the electrical properties of zinc oxide films prepared by physical vapour deposition have been published (Webb and co-workers, 1981; Caporaletti, 1982; Ito and Nakazawa, 1983; Nanto and co-workers, 1984; Vasanelli and co-workers, 1987) which show that low resistivity, infrared reflecting films can be prepared. The methods of deposition used include magnetron and diode systems with zinc targets and deposition from zinc oxide targets in Argon and Argon/Hydrogen atmospheres. Substrate bias has also been employed.

A recent paper by Brett and Parsons (1986) describes the preparation of a zinc black layer overcoated with a zinc oxide heat mirror to produce a good solar absorber with low emissivity. The zinc and zinc oxide layers were produced by controlling the substrate bias in an RF reactive gas system.

Results for the production of both transparent and black zinc oxide films are presented in this paper formed from metal and metal oxide targets with the application of substrate bias. Optical electrical and microstructural properties of these films have been examined.

EXPERIMENTAL

Deposition of zinc oxide films was carried out on a Polaron PT7440 rf sputter coater. Films were sputtered from both zinc metal and zinc oxide targets of 100 mm diameter onto ultrasonically cleaned microslide glass. The zinc metal target was sputtered in Argon/Oxygen reactive gas mixtures whilst the oxide target used Argon/Hydrogen mixtures. Gas mixing was achieved using two Brooks gas flow meters controlled from a power supply/control unit. Total gas pressure was varied between 2×10^{-4} mb and 1×10^{-2} mb. The substrates were watercooled and set 50 mm distance from the target. No intentional substrate heating was used. A bias voltage of up to -200V could be applied to the substrate. The rf forward power could be varied up to a maximum of 500 Watts with the rf matching network tuned to give a reflected power of less than 5 Watts.

A Beckman 5240 UV/VIS/NIR spectrophotometer with an integrating sphere attachment was used to measure the spectral transmittance and reflectance of the films. A 20 point selected ordinate calculation for Air Mass 2 (Wiebelt and Henderson, 1979) was used to determine the solar optical properties. Specular reflectance in the infrared was measured using a Perkin-Elmer 683 spectrophotometer. Electrical properties were measured using a signatron 4-point probe and film thickness determined using a Talystep 4 or from weight gain measurements. Surface microstructure was examined using a Jeol JSM 840 scanning electron microscope. X-ray diffraction studies were performed on a Phillips PW 1720 x-ray diffractometer.

RESULTS

The films that have been produced can be classified into one of three groups. These are: transparent zinc oxide, dark zinc oxide and black zinc oxide coatings.

Transparent Zinc Oxide Coatings

Table 1 summarises the range of conditions that can be used to produce transparent zinc oxide films together with their optical and electrical properties. The spectral transmittance and reflectance curves for two typical films are shown in Figure 1 whilst Figure 2 shows a secondary electron micrograph of one of the coatings.

Dark Zinc Oxide Coatings

Table 2 summarises the range of conditions that were found to produce dark zinc oxide coatings together with their optical and electrical properties.

Black Zinc Oxide Coatings

Table 3 shows the conditions under which black zinc oxide films were made and includes their optical properties. The spectral reflectance curves for three of these films are shown in Figure 3. The surface microstructure of as-sputtered metallic zinc and black oxide films are shown in Figures 4 and 5.

DISCUSSION

The results show that a range of properties of zinc oxide can be obtained using an rf diode sputtering system.

The results show that transparent films can be produced from both the zinc oxide and zinc targets. Whilst their solar transmittance is high (>80%) the electrical conductivity of these films is too low to give the required high infrared reflectance needed for a heat mirror coating. Indeed the sum of the solar transmittance and reflectance for these films on glass substrates is about 0.97 indicating that no appreciable free carrier absorption occurs.

The results for the dark transparent films show that lower resistivity films can be made. However the electrical conductivity is still too low for heat mirror films, an increase of at least 2 orders of magnitude being required. Also the values of solar transmittance are very low. The results of film deposition from the zinc target in Argon/Oxygen atmospheres shows that a transition between metallic absorbing and transparent non-conducting films occurs as oxygen concentration is increased. The dark transparent films can be produced on the point of this transition but reproducibility of these films is not good due to the sharpness of the transition.

The results for the zinc blacks show that high values of solar absorptance can be obtained by a number of methods. The solar absorptance of the as-sputtered metal is high due to the surface roughness of the film (see Figure 4). When a small amount of oxygen is added to the reactive gas higher values of solar absorptance are obtained. This is due to the formation of zinc particles in zinc oxide. The oxide layer acts as an antireflection coating thus increasing solar absorptance.

The surface microstructure of these films differs from that of the as-sputtered metal as can be seen by Figures 4 and 5. These films were sputtered onto glass substrates and appear mirror like when viewed from the back surface.

Another method of achieving a high value of solar absorptance is to overcoat the metal layer with a zinc oxide layer which acts as an antireflection coating.

Both of these types of black zinc oxide coatings can easily be deteriorated by brushing the surface. However they may have some application in evacuated tube solar collectors where degradation is not as great a problem.

The results reported here taken together with those published by others on electrically conducting films indicate that zinc oxide is a candidate material for a range of solar coating applications.

REFERENCES

Brett, MJ, Parsons, RR and Baltes, HP (1986) Zinc oxide multilayers for Solar Collector coatings. Applied Optics 25, 2712 - 2714.

Caporaletti, O. (1982) Electrical and Optical properties of bias sputtered ZnO thin films. Solar Energy Materials 7, 65 - 73.

Ito, K. and Nakazawa, T (1983) Transparent and highly conducting films of ZnO prepared by rf sputtering.Jap. J of Applied Physics 22, L245-L247.

Lampert, CM (1981). Heat mirror coatings for energy conserving windows. Solar Energy Materials 6, 1-41.

Nanto, H, Minami, T, Shooji, S, and Takata, S. (1984). Electrical and Optical properties of zinc oxide thin films prepared by rf Magnetron sputtering for transparent electrode applications. J of Applied Physics 55, 1029-1034.

Vasanelli, L, Valentini, A, and Losacco, A (1987) Preparation of transparent conducting zinc oxide films by reactive sputtering. Solar Energy Materials 16, 91-102.

Webb, JB, Williams, DF and Buchanan, M (1981) Transparent and highly conductive films of ZnO prepared by rf reactive magnetron sputtering. Applied Physics letters 39, 640-642.

Wiebelt JA and Henderson, JB (1979). Selected ordinates for total solar radiant property evaluation. J Heat Transfer 101,-107.

ACKNOWLEDGEMENT

This work was supported in part by the Science and Engineering Research Council.

* System set up altered

Sample	Target	RF Power (W)	Time (min)	Plasma	RGC %	Pressure x10 (mbar)	Bias (V)	τ_s	ρ_s	t (nm)	ρ_o (Ωm)
Aa3	ZnO	50	180	Ar	-	20	-	0.84	0.13	390	12.0
Bc1	ZnO	150	60	Ar	-	10	-	0.81	0.14	550	2×10^{-2}
Zn/6	Zn	150	30	Ar/O_2	20.0	2	50	0.83	0.15	380	6×10^{-2}
Zn/8	Zn	150	30	Ar/O_2	12.5	2	50	0.82	0.15	525	5×10^{-2}
ZnO/9	ZnO	150	30	Ar/H_2	30.1	2	60	0.81	0.17	200	1×10^{-1}
ZnO/10	ZnO	300	30	Ar/H_2	5.2	2	100	0.82	0.15	270	2.7
Zn/02/9*	Zn	150	30	Ar/O_2	5.2	5	60	0.85	0.15	360	4×10^{-2}
Zn/02/21*	Zn	150	30	Ar/O_2	3.7	5	60	0.83	0.15	330	3×10^{-2}

Table 1   The optical and electrical properties of transparent zinc oxide films prepared under differing conditions.

Sample	RF Power (W)	Time (mins)	O_2 Conc (%)	Pressure x10^{-3} (mb)	Bias (v)	τ_s	ρ_s	Sheet Resistivity R ($\Omega/\square$) x 10^3	t (nm)	ρ_o (Ωm)
Zn/02/3	150	30	2.5	1	90	0.020	0.255	3.0	280	8.4×10^{-4}
Zn/02/6	150	30	7.4	1	90	0.022	0.246	5.9	290	1.7×10^{-3}
Zn/02/16	150	30	3.7	5	75	0.035	0.194	9.6	490	4.7×10^{-3}

Diffusion pump baffle position altered before deposition of each film

Table 2   The optical and electrical properties of dark zinc oxide films.

Sample	Target	RF Power (W)	Time (min)	Plasma	Pressure (x10^{-3} mb)	Bias (v)	α_s
Zn/ZnO/4	Zn / Zn	150 / 150	60 / 60	Ar / Ar/O_2(34%)	1 / 1	50 / 50	- / 0.89
Zn/2	Zn	150	60	Ar/O_2(5.6%)	1	50	0.94
Zn/300/16h)	Zn	150	60	Ar — Thermally Oxidised in air at 300°C for 16 hours	1	- / 0.95	0.72
Zn/02/7	Zn	150	30	Ar/O_2(6.6%)	1	90	0.85
Zn/02/19	Zn	150	30	Ar/O_2(4.1%)	2.5	80	0.83
Zn/02/22	Zn	150	30	Ar/O_2(2.1%)	5	75	0.92
Zn/02/27	Zn	150	30	Ar/O_2(3.0%)	5	75	0.95
Zn/02/30	Zn	150	30	Ar	1	75	0.75

Table 3   The optical properties of zinc black absorbers prepared under differing conditions.

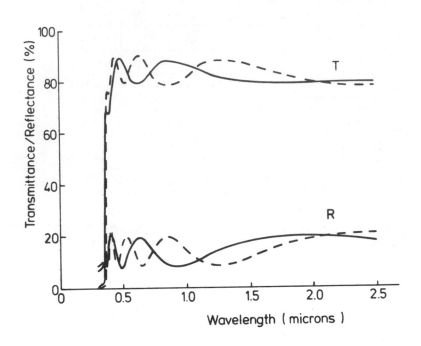

Figure 1   The spectral transmittance (T) and reflectance (R)
            of two transparent films, Zn/6 (————) and
            ZnO/10 (----).

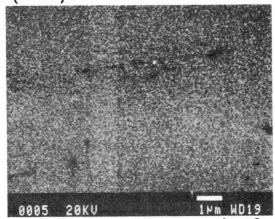

Figure 2   Secondary electron micrograph of a transparent
            zinc oxide film.

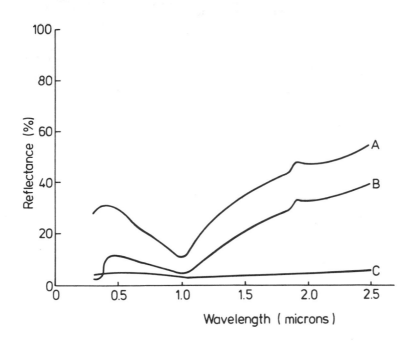

Figure 3   Spectral reflectance curves for three zinc blacks.
            Films Zn/02/03 (curve A), Zn/02/27 (curve C).

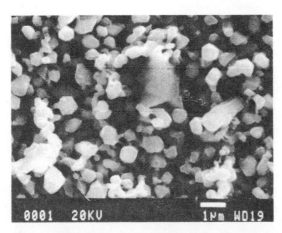

Figure 4   Secondary electron micrograph of the as-sputtered
            metallic zinc film (Zn/02/30).

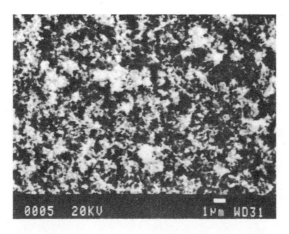

Figure 5   Secondary electron micrograph of a black zinc oxide
            film (Zn/02/27).

OPTICAL PROPERTY MEASUREMENTS ON ADVANCED GLAZINGS

J.L. CHEVALIER

Service Materiaux
Centre Scientifique et Technique du Bâtiment
24 rue Joseph Fourier
38400 SAINT MARTIN D'HERES - France

Tél. : (33) 76 54 11 63 - Télex : 980 149 F

INTRODUCTION

The development of new glazing materials, by introducing more sophisticated products in the building field, araises the question of the optical properties. We shall express here the user's point of view, that is to say the need for tools to identify and to qualify these new materials.

Until now, to determine the performances of the traditional gla-zings, a set of calculation procedures were available, in which well known standard values were used for the optical factors.

The innovative character of the new coated glasses is precisely their optical factors, and measurements are now needed to take it into account in the functional characteristics. The right procedure to perform this measurement is first to identify the relevant factor, and second to measure it using a suitable tech-nique.

In the following development, a special emphasis is given to the emittance measurement, where lies right now the essential discus-sion.

GLAZING CASE ANALYSIS = FROM OPTICAL FACTOR TO FUNCTIONAL CHARACTERISTIC

Functional characteristics

Figure 1 illustrates the three parameters characterizing the various functions of a glazing, as presented recently [1]. They are : the daylight transmittance T_L, the solar factor S, and the U value.

The need of artificial illumination in buildings depends on the daylight transmittance of glazings while air conditioning costs in summer and heating costs in winter are reduced with suitable solar factor and U value.

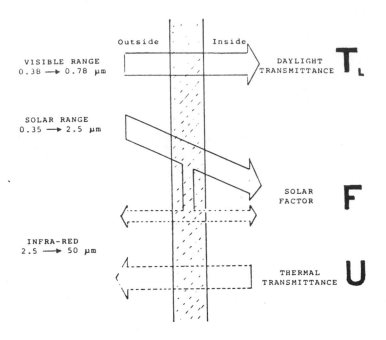

Figure 1 - The functional characteristics of a glazing
and the spectral ranges concerned

The daylight transmittance represents the fraction of natural daylighting penetrating inside through the glazing.

It is obtained by the following formula :

$$T_L = \frac{\int_{380nm}^{780nm} \boxed{\mathcal{T}_\lambda} \cdot W_\lambda \cdot d\lambda}{\int_{380nm}^{780nm} W_\lambda \cdot d\lambda} \qquad (1)$$

Where $W_\lambda = D_\lambda \cdot V_\lambda$, D_λ is the spectral distribution of the illuminant D65 and V_λ the spectral distribution of the human eye sensitivity [2] .

The solar factor or total solar energy transmittance is the sum of the direct solar transmittance $\mathcal{T}_e$ and the secondary heat transfer coefficient, qi (i.e. the fraction of the absorbed energy flowing inwards by convection and radiation).

The formula is given in an ISO draft proposal [3]

$$F = \mathcal{T}_e + qi$$

$$\tau_e = \frac{\int_{300nm}^{2100nm} \boxed{\tau_\lambda}\, s_\lambda\, d\lambda}{\int_{300nm}^{2100nm} s_\lambda\, d\lambda}$$

(S is the solar spectral distribution at the see level)

$$qi = \alpha_e \frac{\alpha_i}{\alpha_o - \alpha_i} \qquad (2)$$

$$\alpha_e + \tau_e + \rho_e = 1$$

$$\alpha_o = C\,(W.m^{-2}\,K^{-1})$$

$$\alpha_i = A + B\,\boxed{\mathcal{E}_i}\,(W.m^{-2}\,K^{-1})$$

For a double glazing, $\tau_\lambda = \dfrac{\boxed{\tau_{1\lambda}} \cdot \boxed{\tau_{2\lambda}}}{1 - \boxed{\rho_{1\lambda}} \cdot \boxed{\rho_{2\lambda}}}$

where $\tau_{i\lambda}$ and $\rho_{i\lambda}$ are respectively the transmission and reflexion factors of each sheet of glass.

The U value or coefficient of thermal transmittance of the glazing is defined as the heat quantity flowing under steady conditions in unit time through a unit surface for each degree of temperature difference between inside and outside. U value is expressed in $W/m^2 °K$, and the formula (3) is given in another draft proposal [4]

$$\frac{1}{U} = \frac{1}{he} + \frac{1}{ht} + \frac{1}{hi}$$

$$\frac{1}{ht} = \sum_{1}^{N} \frac{1}{hs} + d.r \qquad (3)$$

$$hs = hr + hg$$

$$hr = 4\,\sigma\,Tm^3 \left[\frac{1}{\boxed{\mathcal{E}1}} + \frac{1}{\boxed{\mathcal{E}2}} - 1\right]^{-1}$$

The above formula are shown here only to emphasize the need for optical factors values in the calculation. The optical factors needed appear in the small squares and one can see that transmission, reflexion and emission factors are concerned.

Optical factors

These measurements must be performed keeping in mind the objective of calculation. For that we recommand the following procedure [5]:

    - Identification of the relevant optical factor to be measured, i.e. selection of the right factor (reflexion, absorption, transmission, emission) its spectral range (UV, visible range, solar range, IR) and the geometry of the incident, reflected, transmitted or emitted radiation (normal, directionnal, specular or hemispherical).

    - Choice of a suitable measurement technique for the purpose, i.e. measuring the relevant factor, even if the accuracy is not excellent on a scientific level.

    - Calibration of this technique by using periodically a reference device, to perform the same measurement or to measure the reference surfaces used commonly.

To achieve the first step of the above procedure in the case of glazing, the relevant optical factors are identified as follows :

    a - To calculate T_L with the formula (1) we need to measure the spectral transmission factor in all the visible range (from 380 to 780 nm). For daylight transmission the normal factor is convenient (i.e. measured with a normal incident beam and by detection of the normally transmitted beam). Nevertheless, in the case of a special material the transmittance factor at 30 degrees incidence may be needed, and one must consider if the classical correlations [6] are valid for the concerned material. In the case of a diffusing material, it must be taken into account the large difference one can expect between a normal detection and an hemispherical detection. The choice between these two alternatives is not clearly recommanded in the standards and the consequences may be important [7].

    b - For the solar factor (formula (2)), transmission factor in needed in all the solar spectral range (i.e. from 300 to 2100 nm) and although the standard [3] gives no indication on the geometry of the factor, we recommand to measure the normal hemispherical solar transmission factor. In France we calculate also the so called "facteur solaire utile" in which the 30 degrees hemispherical solar transmission factor is needed. In the case of a double glazing, when the reflexion factors of the two sheets are needed, the same geometry and spectral range are required for reflexion and transmission factors.

    c - To calculate the U value with the formula (3) emission factors are needed. The formula is developped in the case of a double glazing, and ε_1, ε_2 are the emission factors of the glass faces in the central area of the glazing. The spectral range concerned is obviously in the infra red part, the right zone being the range corresponding to the emission of the black body at room temperature. For the geometry, the glazing manufacturers have introduced the concept of "effective emissivity" [4], which seems to lie between the normal and hemispherical emission factor, although no scientific justification were given. Considering the thermal aspect and the geometry of a double glazing, it seems more convenient to require the hemispherical emission factor than the normal one, but we shall see in the following chapter that the experimental problems are important.

THE SPECIFIC CASE OF COATED GLASSES

The major innovation in glazing during these last years is obviously the development of coated glass, with an objective of heat loss reduction. The so called low emissivity coating, by reflecting IR radiation, increases the thermal resistance of the glazing. But the goal is to get this result without reducing the daylight transmittance and solar factor. To check correctly the properties of these new glazings, more than ever relevant measurements of the optical factors are expected.

Transmission and reflexion factors

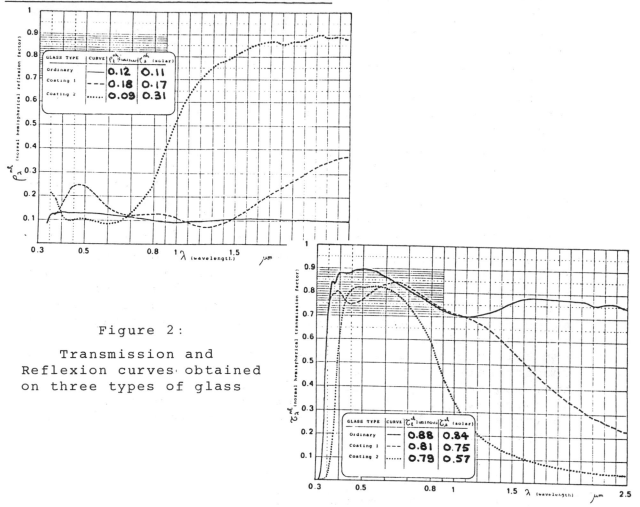

Figure 2:

Transmission and
Reflexion curves obtained
on three types of glass

The figure 2 shows the spectral transmission and reflexion factor
measured on two types of coated glasses, compared to the same curve for a standard float glass. They have been obtained in a
Beckmann 5240 UV-VIS-NIR spectrophotometer equiped with a $BaSO_4$
coated 150 mm integrating sphere. The factor measured is spectral
(in the solar range) and normal hemispherical. The shape of the
curves and the data given illustrate the importance of a good iden-
tification of the relevant optical factor : if a confusion between
luminous and solar factor is acceptable for ordinary glass, it is
essential to distinguish them for coated glasses, due to the major difference between both.

Emission factor

The measurement of the emission factor is actually the most difficult to perform. Several alternative techniques are avalaible [8],
and are illustrated in figure 3.

    a - Calorimetric measurements : by fixing all the physical
        conditions around the sample, the energy provided to main-
        tain it at a given temperature is a function of the hemis-
        pherical emittance of the sample. The equipement needed is

sophisticated, even more for measurement about room temperature.

b - Radiometric measurements : with an IR detector the radiation emitted by the sample is measured and by comparison
with the radiation emitted by a black body at the same temperature, directionnal emittance is determined.
Several measurements are needed at different angles and by
integration hemispherical emittance is obtained but this
method is not valid at room temperature.

c - Broadband devices : these equipements are generally considered as "portable". An IR source acting also as detector
is mounted in a sensor applied on the sample. This sensor
collects the radiation reflected by the sample, and the
energy necessary to maintain the equilibrium temperature of
the source is related to the IR reflectance of the sample.

d - Spectrophotometric measurements : it is possible in a
spectrophotometer to perform measurement of spectral specular reflectance at different angles of incidence using a special accessory. Then by two integrations hemispherical emittance can be reached but this is valid only in the case of
specular surfaces.
For all type of surfaces measurements of emittance in a spectrophotometer is possible using an integrating sphere. But
the spectral range is limited to 20 μm.
A device available in the ISPRA CEC research center is a focusing ellipsoïd : normal hemispherical spectral reflectance
is obtained.

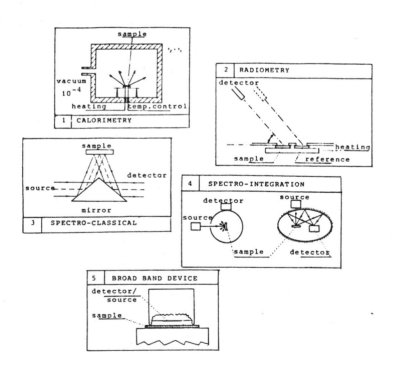

Figure 3 - Existing measurement technique
for the emission factor

All these techniques have been used to measure a commercially available low emissivity coated glass and table I gives the data obtained. The scattering of the values shows the difficulties still existing in this type of measurement to reach a correct result.

According to the draft standard recommandation, the glass manufacturers are used to measure the emission factor by spectrophotometric technique, in the specular near normal mode.

To reach the relevant hemispherical, or "effective" emittance, tables of convenient correlation factors are given, assuming that they are suitable for the type of surface which is measured.

Measurement technique	Factor actually measured	Emittance obtained	Result
Calorimetry	ε_{th}^{h}	ε_{150}^{h}	0.35 ± 0.02
Radiometric technique	ε_{th}^{n}	ε_{100}^{n}	0.38 ± 0.05
	$\varepsilon_{th}^{'}$	ε_{100}^{70D}	0.47 ± 0.05
Classical spectro-photometry	$\rho_{\lambda}^{'}$	ε_{20}^{h}	0.30 ± 0.01
Spectrophotometry with focusing ellipsoïd	ρ_{λ}^{nh}	ε_{100}^{n}	0.37 ± 0.01
Portable device 1	ρ_{th}^{nh}	ε_{70}^{n}	0.036 ± 0.02
Portable device 2	ρ_{th}^{hh}	ε_{80}^{h}	0.34 ± 0.02
Portable device 3	ρ_{th}^{hh}	ε_{170}^{h}	0.35 ± 0.02

h = hemispherical λ = spectral
n = normal   th = integrated on the spectral range corresponding
d = directionnal   to the emission of the black body between 0°
                          and 200°C (if not precised)

Table I : Emittance measurements
of a low emissivity coating on glass

The validity of this assumption has been already discussed many years ago [9], but is still a weakness point of the standard, and the International Commission on Glass is working on this problem. In fact the actual agreed measurement technique is valid only if :
- the surface measured is specular,
- the low giving hemispherical emittance versus normal emittance is valid for the coating concerned,
and these assumption could be a limitation for the characterisation of any new innovative material.

CONCLUSION

To permit a suitable characterisation of advanced glazing materials, measurements of the optical factors are needed. It is necessary to identify the relevant factor to measure, and to use
convenient measurement techniques. Some problems are still without
a satisfactory and universally agreed solution, but comparative
measurements and collaborative work between european laboratories
could be a constructive contribution.

BIBLIOGRAPHY

[1] JL CHEVALIER, P. POLATO "Day light, solar and thermal
parameters of glazings" Workshop on optical properties
measurement techniques - ISPRA (1987)

[2] Publication CIE N.15 (E - 1.3.1.), "Colorimetrie -
recommandations officielles de la Commission Internationale
de l'Eclairage" - (1971)

[3] ISO TC160/SC 2/WG 2, Draft Proposal 9050, "Light transmittance
Direct Solar transmittance, Total Solar Energy transmittance,
Ultraviolet transmittance and related factors of glazings"
- (1984)

[4] ISO TC160/SC 2/WG 2, "Thermal insulation of glazing : calculation rules for determining the steady state U value
(thermal transmittance) of double or multiple glazings"

[5] JL CHEVALIER "Optical measurements in the solar range :
measurement techniques for building surfaces" Workshop on
optical properties measurement techniques - ISPRA (1987)

[6] JR WATERS "Solar heat through unshaded flat glass"- Pilkington laboratory report (1954)

[7] E. KROCHMANN, J. KROCHMANN "Luminous and radiant characteristics of glazing and its measurement" - European Conference
on Architecture - Munich (1987)

[8] JL CHEVALIER et all "Les vitrages à faible émissivité"
22e Colloque CSTB/CFI, Paris, 31 mars 1987, CSTB Magazine
- Mai 1987

[9] JP MILLARD, ER STREED "A comparison of infra red emittance
measurement and measurement techniques" - Applied optica,
vol.8, n°7, p.1485-1492 - Juillet 1979

SUPERGLAZING, U-VALUE=0.9

Paul Robinson, John Littler

Research in Building Group

Polytechnic of Central London

35 Marylebone Road, London

INTRODUCTION

Various calculation techniques have been used to optimise south facing areas of glazing for UK housing, in terms of minimizing energy consumption. In general the optima suggested are about 10% (of the floor area) for single glazing, 20% for double glazing and 25-30% for triple glazing. With the advent of low-e coatings and the possibility of aerogel glazing systems, RIB was interested in documenting the performance in use of very low U-value glazing systems in the UK climate.

Calculations [1] using the thermal simulation model SERI-RES [2] indicated that a glazing of low U-value (approx. 0.8 W/sqmK) and even relatively poor transmittance (shading coefficient=0.5) could with advantage be used on the whole of the south facade of UK houses.

Demonstration support from the EEC (DG XVII) allowed the construction of five superglazed houses and three double glazed controls, at Energy World, Milton Keynes. All houses are being monitored until 1989. The Department of Energy has funded measurements in the US (at Lawrence Berkeley Laboratory, California and Tait Solar, Arizona), and the UK (in situ tests by heat flow mat and polystyrene slab methods) to characterise the performance of the superglazing alone. These measurements and their comparison with predictions from two glazing simulation models are reported here.

MEASUREMENTS ON THE SUPERGLAZING

The superglazing was evaluated using four different experimental techniques.
1. Measurements on a sample sent to the Lawrence Berkeley Laboratory's (LBL) MoWiTT facility, where a sophisticated test room exposes the glazing to outside conditions.
2. Lab. based tests: guarded hot plate U-value tests by Tait Solar, and shading coefficient tests by DSET Labs, Arizona.
3. Heat flow sensors attached to the interior side of the glazing.
4. Temperature difference measurements across a polystyrene slab offered up to the inside of the glazing.

The glazing systems being tested are variations on the "superwindow" advanced glazing system used in the "Courtyard" passive solar house design. The superglazing, shown in Figure 1, consists of a sealed triple glazed unit with two Interpane IPLUS Neutral soft low-e coatings, and two 12mm cavities filled with argon, a low-e louvred blind, and a fourth, protective, outer sash. All components are housed in a Nor-Dan timber frame. Measurements have been made on the system with blinds closed, blinds horizontal, blinds drawn up, and the triple glazed unit alone. An opening double glazed unit, set in a timber frame was also tested in the US.

The results for each assessment technique considered are shown in Table 1.

1. Exposure to Real Weather: MoWiTT, LBL, California.

The MoWiTT (Mobile Window Thermal Test Facility) was stationed at Reno, Nevada, facing south, during testing in February 1987. MoWiTT aims to provide accurate, full-scale, dynamic thermal performance testing of various types of window systems exposed to real weather conditions, including solar gain [3]. Two identical cells are positioned side by side for simultaneous, comparative studies on different types of windows and management strategies. The facility is mobile to vary location (ie. climate) and orientation.

After realism, the key priority of MoWiTT is to make direct calorimetric determinations of heat flow rates through glazing samples as a function of time. This aim is achieved by:
-measuring the heating and cooling required to maintain a constant cell temperature.
-decoupling cell heat loss from ambient temperature with a sur-rounding guard-air plenum maintained at cell temperature.
-completely covering the chamber walls with high sensitivity heat flow sensors. Under solar gain conditions, maintaining equal temperatures in the test cell air and the guard air is not sufficient to prevent heat transfer since interior surface temperatures of the cell will rise above the zone temperature. The thermal storage is thus quantified and an accurate dynamic net heat balance can be determined.
-logging data down to 30 second intervals.

There are two techniques for determining shading coefficients with MoWiTT. One compares heat flow through the test window with the single glazed reference window, housed in the adjacent cell. Another takes the measured test window heat flow profiles, and plots calculated profiles for shading coefficients of 1 and 0.5, with the same environmental conditions. Linear interpolation between the two calculated profiles to reproduce the measured profile gives the shading coefficient of the test window. Continuously clear days are best used for shading coefficient assessment.

2. Laboratory Based Measurements.

Tait Solar Labs., Tempe, AZ. employ a Guarded Hot Plate tech-nique which uses the laboratory air on the cold side, and a

guarded hot plate on the hot side of the window sample. A
perimeter guard encourages heat flow to occur only perpendicularly
through the glass. Since the room is the cold side, the U-value
is assessed at unrealistically high temperatures.

DSET Labs Phoenix, AZ. make lab. based shading coefficient
determinations. They use a solar simulator and calorimeter to
measure the net energy gain of a window system when the inside and
ambient temperatures are approximately equal. The calorimeter
gives the solar heat gain coefficient, F; the shading coefficient
is given by the ratio of F for the test sample to F for a reference single pane (0.87 for ASHRAE conditions [4]). The solar irradiance simulator comprises an array of 55 compact source iodide
lamps in air cooled troughs. For intercomparison of results,
tests are carried out under ASHRAE Summer environmental
conditions, ie. a solar irradiance of 782 W/m^2, room temperature
at 24°C, and a windspeed of 3.4m/s. The simulated wind hits the
sample at a 60° angle of incidence, and the simulated radiation
at 10°.

In Situ U-value Measurements.

3. Heat Flow Mat Technique.

EMC made measurements [5] using commercial heat flux sensors
mounted on the glazing in December 1987. Two heat flux sensors
were used to measure windows in different configurations side by
side (eg. blinds open versus blinds closed for identical environmental conditions). Radiation shielded thermistors are used.

The thin flux sensor slightly affects local thermal resistance,
and the heat flux is changed relative to an undisturbed glass
surface; measurements are therefore unrepresentative (this effect
reduces in magnitude as the resistance of the glazing increases).
The local changes in thermal resistance results in multi-
dimensional heat flow paths; this causes errors as the design and
calibration of heat flow sensors are based on one dimensional heat
flows. The local convective heat transfer coefficient is changed
by the presence of the heat flow sensors.

In the EMC report, correction factors are suggested for these
three problems.

To assess edge effects the heat flow sensors were both positioned
on the same window, one in the centre and one 30mm from the mid-
point of a vertical edge. In order to produce a figure comparable
with MoWiTT U-values for the glazing and frame, from the heat flow
mat measurements, it is necessary to estimate the effect of the
different U-value at the edge of the glazed area for a window of
the same size, and to put in a frame correction. The assumption
is made that the heat flux measured 30mm from the edge of the
window occurs uniformly across a 60mm wide band around the edge
of the window. Inside this area the heat flux is assumed to be
as measured at the centre. The heat flux through the frame is
calculated from the U-valves of frame materials and thicknesses
documented in the Norwegian Standard NS 3031 [6].

For the triple glazed unit, the degradation of the U-value-the difference between centre of pane U-value and adjusted value arising from edge frame effects - is 29%. For the superglazed unit (blinds closed) there is a 23% degradation in U-value due to edge and frame effects.

4. Polystyrene Slab Technique. [7]

This method involves covering the test window with an accurately cut polystyrene slab of the same size and known thermal transmissivity. When a cover is added to the window, for steady state conditions, the rate of heat loss through the covering is constant and equal to the heat loss through the window. Assuming one dimensional heat flow at the centre of the pane, the U-value of the test window (neglecting its inside surface heat transfer coefficient) can be found from steady state measurements of the temperature difference between the room-side cover surface and the cover/window junction, and the cover/window junction and outside air temperature; the thickness of the cover and its thermal transmissivity being known.

A thickness of polystyrene is chosen which makes the thermal resistance of the cover and window roughly equal, thus minimizing errors and the time taken to reach equilibrium. To prevent air-flow between cover and window and formation of an extra air gap, the cover must be taped to the window frame and the polystyrene pressed against the window. Using this technique, physics students from the University of Bath attempted measurements on an east facing superglazed unit without blinds, in a Courtyard house, in February 1987.

Adjustment of these results to represent the U-value of the sample size tested in MoWiTT, for intercomparison, proceeds in the same way as described for the heat flow mat method.

GLAZING SIMULATION MODELS. WINDOW 2.0 AND MULTB.

WINDOW 2.0 is a computer program for calculating U-values and shading coefficients of glazing systems described by the user, for environmental conditions specified by the user. It also calculates glazing layer temperatures and heat flux through the windows. The program was developed by the Windows and Daylighting Group of the Lawrence Berkeley Laboratory in California [8].

WINDOW 2.0 performs the U-value calculation by assuming a one dimensional temperature profile across the centre of a glazing configuration. Radiative treatment is by 2-band modelling: thermal and shortwave, with fundamental radiative theory being used to calculate exchange between panes. Convective heat transfer is calculated using semi-empirical correlations. The temperature distribution is refined iteratively by a finite difference method until the required accuracy is obtained. In WINDOW 2.0 the glazing U-value obtained is for an infinite window, and the frame corrected U-value assumes this glazing U-value right up to the boundary of the glass with the frame. Overall light transmission of a window and its effective absorption are calculated in WINDOW 2.0 from the transmittance and reflectance of the isolated layers making up the window. The treatment is only approximately valid as polarization and wavelength dependencies are averaged for the properties of the individual layers.

MULTB was developed by Pilkington Glass LImited [9].
The program computes solar and visible optical properties,
shading coefficients, glass and backup temperatures, and basic
temperature differences for a multiple glazing system, one layer
of which can be a blind. A thermal safety check can be made on
each layer.

MULTB is also a steady state model, and works in a similar way
to WINDOW 2.0. The frame correction treatment is more sophisticated
than that used by WINDOW 2.0. Measured shading coefficients are
affected by the overshadowing of the glass by the frame, at
oblique sun angles, if the glass is recessed. The two glazing
simulation models do not take account of this. Work at PCL on
test cells has shown how serious this can be [10].

Measurements of the properties of Interpane IPLUS Neutral Low-e
Coated Glass used as input to the two progrms, was made by the
Energy Equipment Testing Service, Cardiff.

COMMENTS ON RESULTS

Table 1. shows results from all the assessments performed on
the superglazing in its various configurations, and the double
glazed unit. Where measurements have been made on samples other
than those used in the MoWiTT tests, then the figures have been
adapted for comparison.

The in situ technique of U-value measurement using heat flow mats,
applied by EMC, gives results which compare very well with other
assessment methods. It is a relatively cheap and unobtrusive
assessment method for investigating windows in existing buildings.
The Polystyrene Slab technique gives a low value for the superglazed configuration with no blinds, compared with the MoWiTT
assessment; no provision was made to avoid an air gap between
the slab and the window, which probably caused this low value.
The technique is an easy and cheap method of U-value assessment.
The technique would benefit from further refinement and
development.

The Hot Plate measurements were made at temperatures more in line
with the ASHRAE Summer Standard conditions, than the winter
conditions under which the other tests were conducted, and which
were used to drive the models. This could explain the
discrepancy between some of the Hot Plate measurements and the
other assessments. The Hot Plate equipment is fairly low cost
and simple in comparison with the MoWiTT, but does seem to give
reasonable assessments for even the advanced glazing systems
under investigation here; the triple glazed unit being the only
system not reasonably evaluated. Relative assessments can be
easily, and from the evidence here, accurately made. A
discrepancy in the MoWiTT results is that the U-value for the
triple glazed unit, 1.13+/-0.13, is lower than that for the
superglazing with blinds pulled up, 1.19+/-0.12. The errors
associated with each value indicate that there is no measurable
difference between the two configurations. U-value is a

35.

parameter subject to some variation, depending on environmental conditions, and inspection of the weather data for the two test periods shows much higher winds for the superglazed test with blinds pulled up (5-15mph), than for the triple glazed test (5mph); this probably explains the discrepency. The intercomparison between the models and the comparison of model predictions with the experimental data from the MoWiTT tests, suggest a high level of consistency of the models, and confidence in the predictions they make on window performance.

Simulation of any but the triple glazed of the superwindow configurations cannot be done accurately as both the models assume all cavities are sealed, when, on the actual window the fourth pane forms a vented cavity; this is the reason for the low assessment by WINDOW 2.0. The shading coefficient calcula- tion is not invalidated by this consideration.

REFERENCES.

1. Littler J, "Design Considerations for Solar Courtyard Houses" Sun at Work in Europe, Oct. 1986 p10
2. Haves P, Littler J, Refinements to SERI-RES. Dept of Energy. ETSU S-1130 1987.
3. Klems J H, 1984. "Measurement of Fenestration Performance Under Realistic Conditions". Proceedings of Windows in Building Design and Maintenance conference, Gothenburg.
4. ASHRAE Handbook 1981 Fundamentals. American Society of Heating, Refrigeration, and Air-Conditioning Engineers Inc. Atlanta, GA.
5. Energy Monitoring Company report for the RIB Group, Dec 1987. "In situ U-value measurements on high performance glazing at the Polytechnic of Central London Courtyard Houses".
6. Norwegian Standard NS 3031.
7. Swane J R, Final Year Project Report for Bath University BSc course in Applied Physics, May 1987. "Aspects of Energy Design in a House Heated by Passive Solar Power".
8. Arashteh D. et al, 1986. "Window 2.0 User Reference Guide". Windows and Daylighting Group, Lawrence Berkeley Labs.
9. Personal Communication from P G T Owens, Environmental Advisory Service, Pilkington Glass, St Helens.
10. Littler J and Walker J, 1987. SERC Final Report on Validation.

Figure 1   Superwindow design, comprising a sealed triple glazed unit
with two Interpane IPLUS Neutral soft low-e coatings and
12mm cavities filled with argon gas, a low-e louvred blind
and a fourth outer pane. All components are housed in a
Nor-Dan timber frame.

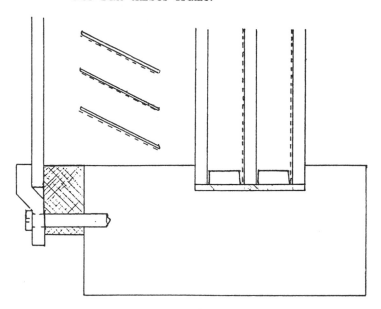

TABLE 1. COMPARISON OF RESULTS FROM VARIOUS ASSESSMENT METHODS USED.

	S.G.blinds closed	S.G.blinds open	S.G.blinds pulled up	Triple 2 low-e	Double
MoWiTT Field U.	0.89 +/-0.13	1.03 +/-0.20	1.19 +/-0.12	1.13 +/-0.13	2.41 +/-0.39
Field S.C.	0.26	0.22	0.48	0.62	0.80
Tait Solar Hot Plate, lab. U.	1.08	-	1.25	1.42	2.38
DSET Solar Simulator S.C.	0.30	-	0.48	0.58	-
EMC in situ Heat Flow Mat U.	0.95 +/-0.07	-	-	1.07 +/-0.07	-
Poly. Slab in situ U.	-	-	0.84 +/-0.09	-	-
WINDOW 2.0 simulated U.	-	-	0.98	1.10* +/-0.01	2.47* +/-.07
simulated S.C.	-	-	0.45	0.52*	0.87*
MULTB simulated U.	0.96	-	-	1.14* +/-0.03	2.55* +/-0.17
simulated S.C.	0.13	-	0.47	0.52*	0.84*

NB. Simulation assessments are with frame corrections.

KEY: * = simulation driven by MoWiTT data recorded during corresponding
test. Otherwise, simulations are for ASHRAE Winter Conditions.

SC = Shading Coefficient,   U = U-value.

37.

FORWARD SCATTERING OF INSOLATION THROUGH TRANSPARENT AND TRANSLUCENT MATERIALS

B. Norton, S.A.M. Burek, C.J.B. Girod, D.E. Prapas & S.D. Probert

Solar Energy Technology Centre
School of Mechanical Engineering
Cranfield Institute of Technology
Bedford MK43 OAL, U.K.

ABSTRACT

An experimental technique for determining the forward scattering of the direct beam during transmission through solar optical materials under either outdoor or simulated indoor insolation conditions is described.

INTRODUCTION

The direct component of incident insolation may be forward scattered during its passage through a transparent or translucent media due to (i) inhomogeneity of the material, this is most noticeable in plastics, and (ii) dust and/or condensation surface deposits.

To predict accurately the behaviour of many solar energy systems it is essential to quantify such forward scattering or "haze". To demonstrate this, consider the optical behaviour of a parabolic-trough concentrating (PTC) solar-energy collector. During passage through the transparent aperture cover, insolation suffers both transmission losses and forward scattering. The latter may be increased by dust deposited on the cover. The optical performance of the PTC is impaired by the subsequent deterioration of the collimation of the direct insolation. The forward scattering coefficient, σ, denotes that fraction of the direct insolation forward scattered though the aperture cover. An illustrative value of $\sigma = 0.03$ has been used in a numerical ray-tracing analysis [1] to illustrate the effect of scattering on the optical efficiency of a PTC collector. As can be seen in Fig. 1, the optical efficiency is reduced when scattering occurs.

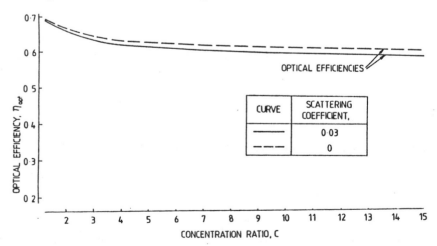

CURVE	SCATTERING COEFFICIENT,
——	0·03
– – –	0

Fig. 1.

BACKGROUND

Standards for determining haze utilise an 'integrating sphere': transmitted light is reflected around the matt, but highly reflective, internal surface of the almost complete sphere, and its intensity orthogonal to the entrance port, which is covered by a sample of the material in question, is measured with a photoreceptor [2,3]. Spectral transmittance data, for several materials measured by methods such as these have been published [4-8]. Similar methods using integrating spheres are described for solar-energy applications in ASTM standards [9-10], and have been the bases of solar transmittance properties [11, 12].

The specimen dimensions, where stated, used in standard procedures, are however insufficient for measurements on non-isotropic and non-uniform (e.g. corrugated or woven) materials. A 2.4 m diameter integrating sphere, with a 300 mm diameter aperture has been built to overcome this [13]. However, although standards account for haze (i.e. forward scattering) measurements, baffles were used to prevent light scattered during transmission from impinging on the photoreceptor.

ASTM standard E242 'Method B' [9] determines transmittance to total solar radiation with the plane of the test material perpendicular to the sun's rays. Several investigators have used a modified version of this method to obtain the variation of transmittance with incidence angle [14-18]. The test conditions are specified as a 'clear, sunny day' which, although implying a low proportion of diffuse insolation, are less than rigorous. Plastics degrade with duration of exposure to environmental influences, and the concomittant reductions in transmittance have been studied for selected materials [19-21]. The present study considers the forward scattering of the direct component of insolation during transmission through transparent and semi-transparent materials. In addition the additional forward scattering introduced by surface deposits and condensation is quantified for representative cases. Previously, only the reduction in transmittance arising therefrom, has been considered [22-24], and for dust deposits, in relation only to exposure time in particular climates.

ANALYSIS

To determine the magnitudes of its diffuse and direct components, the global insolation and its diffuse component are measured simultaneously, the direct component being the difference between the two, i.e.

$$I_b = I_t - I_d \qquad\qquad (1)$$

Similarly, for the total measured insolation transmitted through a material, i.e.

$$S_b = S_t - S_d \qquad\qquad (2)$$

A fraction, σ, of the external beam insolation is forward scattered during transmission. The measured diffuse transmitted insolation will thus include a scattered beam insolation component, i.e.

$$S_d' = S_d + \sigma S_b \qquad (3)$$

For the beam insolation, transmittance terms are defined for the unscattered and the scattered components, i.e.,

$$\tau_{bu} = S_b(1 - \sigma)/I_b \qquad (4)$$

and

$$\tau_{bs} = \sigma S_b/I_b \qquad (5)$$

Hence, the total beam transmittance can be defined as:

$$\tau_b = \tau_{bu} + \tau_{bs} \qquad (6)$$

which is equivalent to:

$$\tau_b = S_b/I_b \qquad (7)$$

The 'true' and measured 'apparent' diffuse transmittance values can be defined respectively as

$$\tau_d = S_d/I_d \qquad (8)$$

and

$$\tau_d' = S_d'/I_d \qquad (9)$$

Substituting for S_d, from equation (3), into equation (8), and then from equation (5) for S_b and equation (9) for S_d', gives

$$\tau_d' = \tau_d + \tau_{bs}(I_b/I_d) \qquad (10)$$

Thus, assuming that τ_d and τ_{bs} remain invariant, a plot of experimental data for τ_d' (i.e. S_d'/I_d) against I_b/I_d for a particular material should yield a straight line with slope τ_{bs}. If the experimental error is small, the scatter of the points about the straight-line will indicate the truth of the assumption that τ_{bs} is a constant. When the incident beam insolation is zero, $S_d' = S_d$ and equation (9) reduces to equation (8). Thus the intercept on the ordinate axis of a graph of τ_d' against I_b/I_d gives the value of τ_d. Eliminating S_b from equations (2) and (3), and re-arranging yields

$$\sigma = (S_d' - S_d)/(S_t - S_d) \qquad (11)$$

and then substituting for S_d from equation (8) gives

$$\sigma = (S_d' - \tau_d I_d)/(S_t - \tau_d I_d) \qquad (12)$$

Re-arranging equation (5) and substituting for S_b from equation (7) leads to

$$\sigma = \tau_{bs}/\tau_b \qquad (13)$$

The definition of 'haze' [2], is the ratio of light scattered during transmission to the total light transmitted. Thus as can be seen from equation (13), the forward scattering coefficient, σ, is the equivalent of haze for the solar spectrum.

EXPERIMENTAL PROCEDURE

A box, matt black internally, with a 1.5 m^2 open face and of adjustable inclination was covered with a sample of the material under test. It contained two Moll-Gorczynski pyranometers: to measure the transmitted total and diffuse insolation respectively. Both indoor and outdoor tests were conducted. In the outdoor tests the sample-covered side of the box faced south and was inclined orthogonally to the direct insolation at solar noon. Thus the azimuthal solar motion provided the variation in incidence angle for the beam radiation. Two solarimeters were located nearby, in the same plane as the test surface in order to measure the incident levels of global and diffuse insolations. A shade-ring correction factor was required for the measured values of diffuse insolation: this depends on the reflectance of the ground and the tilt angle of the solarimeter. A simple practical method for determining the shade-ring correction factor was used [25]. The outdoor test data consisted of the date and simultaneous values of I_t, I_d, S_t and S_d', and the time.

The indoor tests were conducted under a solar simulator. In order to vary the angle of incidence of the radiation on the test specimen, the test box was pivoted about its bottom edge. During the indoor tests, the magnitude of the insolation was measured at discrete angles of incidence with the solarimeters in the test box. 'Incident' radiation levels were measured with the same solarimeters with no cover on the open face of the box.

Results for a 'Clean" Material

Figures 2 to 5 show a particular representative example of data for a clean material. Figure 5 presents a plot of τ_d' (i.e. S_d'/I_d) against (I_b/I_d). From equation (10), the slope of the regression line gives the 'scattered transmittance' of the beam radiation τ_{bs}, and the intercept on the ordinate axis is the transmittance to diffuse insolation, τ_d. In most cases, the degree of scatter of such data was small, suggesting that only a small dependence of τ_{bs} on the angle of incidence ensued. A least-squares regression line was fitted to the data, in order to determine the values for τ_d and τ_{bs}. This is also shown in Fig. 8. By substituting for S_b and σS_b from equations (2) and (3) into equation (4), τ_{bu} can be determined from measured data via

$$\tau_{bu} = (S_t - S_d')/I_b \tag{14}$$

Similarly, by substituting for S_b from equation (3) into equation (5), and using the relationship for τ_d in equation (8), τ_{bs} can be determined from the measured data, using the value of τ_d obtained from the regression analysis:

$$\tau_{bs} = (S_d' - \tau_d/I_d)I_b \tag{15}$$

The relationships in equations (14) and (15) are plotted in Figs. 2 and 3. Figure 4 shows a plot of the scattering coefficient against angle of incidence, as calculated from equation (12).

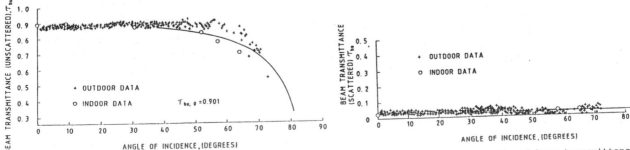

Fig. 2. Variation of the unscattered beam transmittance with incidence angle for the 50 μm-thick polyethylene teraphthalate film.

Fig. 3. Variation of the scattered beam transmittance with incidence angle for the 50 μm-thick polyethylene teraphthalate film.

42.

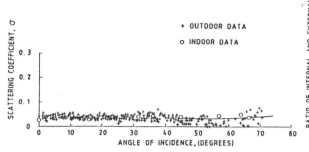

Fig. 4. Variation of the forward-scattering coefficient with incidence angle for the 50 µm-thick polyethylene teraphthalate film.

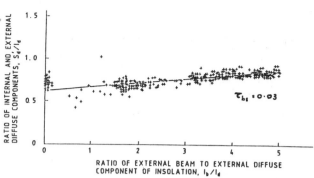

Fig. 5. Graphical representation of equation 10 using date for the 50 µm-thick polyethylene teraphthalate film.

Results for Dusty and Wetted Materials

Surface deposits like dust or condensation enhance two phenomena, reduction of transmittance and forward scattering. The dust used was chosen for the purpose of experiments because its beige colour appeared subjectively to be close to the dust that deposits on solar collectors' covers. This powder was sieved to four different sizes, of diameters 63 to 75 μm, 75 to 90 μm, 90 to 106 μm, and 106 to 125 μm. These were applied to a plane polyethylene surface. For each inclination of the box, the same set of measurements was recorded as with clean surfaces. The scattering coefficient for particles of diameter between 63 and 75 μm is shown in Fig. 6. The results showed that scattering is proportional to the density of coverage by the particles on the surface, regardless to their size.

The effect of water droplets deposited on the inner side of the collector cover was also investigated experimentally. The amount of water sprayed on the cover was determined by weighing the water atomiser before and after application. The quantities used were: 15, 18 and 28 g/m^2. Figure 7 shows scattering coefficient versus the angle of incidence of the beam radiation. Under the conditions of the experiment, the unscattered beam transmittance was found to be linearly dependent on the quantity of water

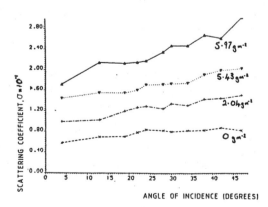

Fig. 6. Scattering coefficients for different surface densities of 63 to 75 µm diameter dust particles on a plane polyethylene film.

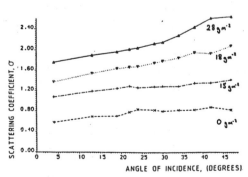

Fig. 7. Scattering coefficient for different surface densities of water sprayed onto a plane polyethylene film.

43.

sprayed on the cover.

ACKNOWLEDGEMENTS

The authors thank various manufacturers, the Science and Engineering Research Council, Swindon, U.K., the Commission of the European Communities, Brussels, Belgium, and the State Scholarship Foundation, Athens, Greece, for supporting the investigations of which this forms a part, financially, or in kind.

NOMENCLATURE

I   Incident insolation (i.e. measured before transmission through the material); W/m^2

σ Forward-scattering coefficient

τ Transmittance

S   Transmitted insolation; W/m^2

S'   Measured transmitted insolation (see equation 3); W/m^2

τ' 'Apparent' Transmittance

Subscripts

0   At normal incidence

b   Of the direct insolation

bs   Of the scattered direct insolation

bu   Of the unscattered direct insolation

d   Of the diffuse insolation

t   Of the total insolation

REFERENCES

1. Prapas, D.E., Norton, B. & Probert, S.D., Optics of Parabolic-Trough Concentrating Solar-Energy Collectors, Possessing Small Concentration Ratios, Solar Energy, 36, 6, 541-550, (1987).

2. Anon, ASTM D1003-61. Standard Test Method for Haze and Luminous Transmittance of Transparent Plastics, American Society for Testing and Materials, Philadelphia, U.S.A., (1979).

3. Anon, BS 2782:1970 Methods of Testing Plastics - Part 5: Method 515A Haze of Film, British Standards Institution, London, U.K. (1970).

4. Trickett, E.S., & Goulden, J.D.S., The Radiation Transmission and Heat Conserving Properties of Glass and Some Plastic Films, J. Agricultural Engineering Research, 3, 281-287, (1958).

5. Bowman, G.E., A Comparison of Greenhouses Covered with Plastic Film and with Glass, Proc. 16th International Horticultural Conference, Brussels, Belgium, 551-448, (1962).

6.   Hanson, K.J., The Radiative Effectiveness of Plastic Films for Greenhouses, J. Applied Meteorology, 2, 793-797, (1963).

7.   Greer, K.R., The Use of Acrylic Sheets for Greenhouses, Acta Horticulturae, 9, 217-220, (1968).

8.   Ratzel, A.C., & Bannerot, R.B., Optimal Material Selection for Flat-Plate Solar-Energy Collectors Utilising Commercially-Available Materials, AIChE Symposium Series, 73, 187-203, (1977).

9.   Anon, ASTM E424-71, Standard Test Methods for Solar-Energy Transmittance and Reflectance (Terrestrial) of Sheet Materials, American Society for Testing and Materials, Philadelphia, U.S.A., (1985).

10. Anon, ASTM E903-82, Standard Test Method for Solar Absorptance, Reflectance, and Transmittance of Materials using Integrating Spheres, American Society for Testing and Materials Philadelphia, U.S.A., (1982).

11. Pettit, R.B., Hemispherical Transmittance Properties of Solar Glazings as a Function of Averaging Procedure and Incident Angle, Solar Energy Materials, 1, 125-140, (1979).

12. Zerlaut, G.A. & Anderson, T.E., Multiple-Integrating Sphere Spectrophotometer for Measuring Absolute Spectral Reflectance and Transmittance, J. Applied Optics, 20, 3797-3804, (1981).

13. Zerlaut, G.A. & Anderson, T.E., A Large Multi-Purpose Solar-Illuminated 8-ft Integrating Sphere, Proc. SPIE Conf. Optical Materials Technology for Energy Efficiency and Solar Energy Conversion (III) San Diego, U.S.A. 152-160, (1984).

14. Edlin, F.E., Plastic Glazings for Solar Energy Absorption Collectors, Solar Energy, 2, 3-6, (1959).

15. Godbey, L.C., Bond, T.E., & Zornig, H.F., Transmission of Solar and Long-Wavelength Energy by Materials used as Covers for Solar Collectors and Greenhouses, Trans. ASAE, 22, 1137-1144, (1979).

16. Robbins, F.V., & Spillman, C.K., Solar Energy Transmission Through Two Transparent Covers, Trans. ASAE, 22, 1224-1231, (1980).

17. Bhaduri, S., & Nguyen, N.H., Transmissivity of Solar Collector Covers, ASME Paper 83-WA/Sol-17, (1983).

18. Fintel, B.W., & Jakubowski, G.S., Obtaining Solar Collector Cover Transmissivities from a Solar Simulator, ASME paper 85-WA/Sol-3, (1985).

19. Rainhart, L.G., & Schimmel, W.P., Effect of Outdoor Ageing on Acrylic Sheet, Solar Energy, 17, 259-264, (1975).

20. Kimball, W.H., & Munir, Z.A., The Effect of Accelerated Weathering on the Degradation of Polymeric Films, Polymer Engineering and Science, 18, 230-237, (1978).

21. Cheng, H., & Bannerot, R.B., On the Weathering of Thin Plastic Films, ASME J. Solar Energy Engineering, 105, 149-156, (1983).

22. Garg, H.P., Effect of Dirt on Transparent Covers in Flat-Plate Solar Energy Collectors, Solar Energy, 15, 299-302, (1974).

23. Sayigh, A.A.M., Al Jandal, S. & Ahmed, H., Dust Effect on Solar Flat Surfaces Devices in Kuwait, Proc. Int. Symposium on Thermal Application of Solar Energy, April 7-10, Hakone, Japan, (1985).

24. Hseih, C.K. & Rajvanshi, A., The Effect of Dropwise Condensation on Glass Solar Properties, Solar Energy, 19, 389-393, (1977).

25. Burek, S.A.M., Norton, B., & Probert, S.D., Analytical and Experimental Methods for Shadow-Band Correction Factors for Solarimeters on Inclined Planes under Isotropically-Diffuse and Overcast Skies, Solar Energy, 40, 2, 151-160, (1988).

ELECTROCHROMIC PROPERTIES OF TUNGSTEN OXIDE FILMS

S.M. Christie and M.G. Hutchins
Solar Energy Materials Research Laboratory,
School of Engineering,
Oxford Polytechnic,
Oxford, England

ABSTRACT

A brief review of the electrochromic properties of tungsten
oxide films for potential application in dynamic window
systems is presented. Requirements for electrochromic cell
constituents are assessed. Experimental results of
crystallinity, colouration efficiency, electrochromic
sensitivity and degradation mechanisms of WO_3 films formed by
reactive rf sputtering and other deposition techniques are
detailed.

INTRODUCTION

An electrochromic material is one which will undergo a
reversible colour change on the application of an external
electric field. This results in a change in the optical
properties of the material, a phenomenon which has led to the
development of electrochromic devices for window and
automobile applications. Much of the early work carried out
on electrochromic materials was in conjunction with optical
displays. With the introduction of light-emitting diodes and
liquid crystals with superior properties, such as faster
switching times, work in this area diminished. More recently
interest has focused on the possibility of incorporating
electrochromic materials as part of a window glazing device.
Present heat mirror coatings, used to regulate solar
transmittance and infrared reflectance, although proving
successful have only static optical properties and hence are
unable to cope with daily or seasonal changes in temperature.
On the other hand, with an electrochromic material it is
possible to control the amount of radiant energy passing
through the window.

A wide range of both organic and inorganic compounds are known
to exhibit electrochromic properties. The inorganic compounds
which include many transition metal oxides, e.g. WO_3 NiO,
V_2O_5, are of particular interest for glazing applications.
The properties of these surfaces have been recently reviewed
(1). In this paper we concentrate only on the preparation and
properties of electrochromic cells employing WO_3 based
electrochromic layers.

REVIEW OF WO$_3$ MULTILAYER STRUCTURES

Most of the published work has focused on WO$_3$ as this was one of the first electrochromic materials to be used in display applications. Electrochromic colouration is exhibited in both amorphous and crystalline forms and occurs by the simultaneous injection of electrons and cations resulting in the formation of a tungsten bronze according to the reaction (2).

$$WO_{3-y} + xe^- + xM+ \rightleftharpoons M_x WO_{3-y} \qquad (1)$$

$$\text{colourless} \qquad\qquad\qquad \text{blue}$$

where $0 < x < 0.5$ and $y < 0.03$. M+ may be H+, Li+ or Na+.

Figure 1 represents a typical electrochromic device that could be used in window designs (3). It consists of two transparent electrodes across which an electric potential can be applied, an electrochromic layer where colouration occurs, an ion storage layer which provides the ions for colouration and finally an ion conductor layer through which the ions pass.

Device designs vary depending upon whether a solid or liquid electrolyte is used. In liquid devices both ion conductor and storage layers are replaced by the electrolyte whereas in a solid state device the ion conductor can be a dielectric and the ion storage layer another electrochromic material.

Transparent Electrodes

It is important that the electrodes are transparent and highly conductive with a low electrical resistance to minimise current loss and heating of the window. Care should be taken to ensure that the electrodes do not react with the electrochromic or electrolyte layers and that the electrode is deposited without causing deterioration of the electrochromic properties of the underlying layers. High performance infrared reflectors like indium tin oxide (ITO) are used as the electrode at the glass interface while semi-transparent metals such as gold have been used as the outer electrode in solid state devices. However, the problem with using a gold electrode, apart from the expense, is that solar transmittance values are greatly reduced. Replacing the layer with ITO will increase transmittance slightly but the conductivity of the film is poor as care must be taken to deposit it at temperatures lower than 135°C otherwise the amorphous WO$_3$ layer will undergo irreversible structural changes (4).

Electrolytes

Work has been carried out using both solid and liquid electrolytes although for window applications a solid would be more desirable. Liquid electrolytes may include a H$_2$SO$_4$ - glycerol mixture providing H+ ions for the formation of the tungstate bronze. However, because water has been shown to limit the lifetime of the WO$_3$ layer, this has led to the use of aprotic electrolytes with ions such as Li+ and Na+. An example of such an electrolyte is LiClO$_4$ in propylene carbonate or ethylene-glycol. Although the mobility of Li+ in Li_xWO_3 is an order of magnitude slower than H+ in H_xWO_3 , Li_xWO_3 is less susceptible to colour fading and is thermodynamically more stable.

Solid state electrolytes may or may not require the presence of water for effective operation. In cells containing LiF or MgF_2 solid electrolytes, the protons needed for the formation of a tungsten bronze are derived from water molecules trapped in the layers during deposition and/or entering from the surrounding environment. On the application of an electric field the water molecules dissociate providing H+ ions for injection into the electrochromic layer (5). A solid electrolyte which does not depend on the presence of water is $Na+ - \beta - Al_2O_3$ (6). The fast ion conductor LiN_3 is another possible electrolyte (7). A new solvent free polymer electrolyte under investigation is poly(ethylene oxide) based polyurethane containing potassium trifluoromethane sulphonate, $KCF_3 SO_3$ (8). In this device in addition to the WO_3 layer there is an electrochromic prussian blue layer which also colours on the same cycle as the WO_3 film.

REPORTED RESULTS FOR WO_3

Results vary depending on method of film preparation, film structure and electrolyte used to induce film colouration. Amorphous films deposited by rf sputtering have been compared with those deposited by vacuum evaporation. Results proved to be comparable with film resistivities of $\sim 10^9$ Ωcm and spectral transmittance values of $\sim 85\%$. The colour and electrochromic properties of the rf deposited films were found to be dependent upon operating pressure and O_2 content (9).

Improved colouration efficiencies have been achieved in films prepared by rf magnetron sputtering when compared with those deposited by vacuum evaporation or ion plating (10). Increases in these efficiencies have been achieved by either increasing the sputtering gas pressure from 8×10^{-4} to 3×10^{-2} Torr or increasing the rf power. Heating of as-deposited films prepared by vacuum evaporation has resulted in a decrease or increase in colouration efficiency depending on annealing temperature, while power and deposition rates have affected the colouration efficiency of films produced by reactive ion plating. These observations have been related to the degree of crystallinity, structure and stoichiometry of the films.

Crystalline films of WO_3 also vary in structure depending on film prepa.ration. Figures 2(a) and 2(b) are x-ray diffraction traces in the $2\theta = 23^\circ - 25^\circ$ range showing the principal peaks in a thermally evaporated and rf sputtered film respectively (11).

It was necessary to crystallise the thermally evaporated film by heating at 400° C for 24 hours. The two traces are markedly different perhaps indicating a preferred orientation in the sputtered film. Quantitative transmission microscopy (TEM) studies on this film revealed a layer-like structure of randomly stacked platelets with their planes parallel to the substrate. The difference in structure affects the electrochromic properties of the films as shown in Figure 3. At 2.5 μm the infrared reflectivity of the sputtered film (solid line) is >60% whereas that of the thermally evaporated film (dashed line) is ~35%. It is believed that the platelets in the rf sputtered films offer less electron scattering and thus higher infrared reflectivity than the crystallised structure of the thermally evaporated films (11).

Although solid state electrolytes would be of more practical use than liquid electrolytes, results obtained using liquids have produced superior results. Amorphous films deposited by reactive electron beam evaporation have been investigated using the liquid electrolyte 1M dehydrated $LiClO_4$ in propylene carbonate and the solid electrolyte MgF_2. The results are shown in Figures 4 and 5 respectively (3).

Evaluation of the integrated solar transmittance has produced values ranging between 86% and 12% for the liquid electrolyte but only between 25% and 3% for the solid. These low transmittance values are due mainly to the reflectance of the gold electrode.

EXPERIMENTAL RESULTS AND DEGRADATION MEASUREMENTS

Our studies of tungsten oxide electrochromic properties commenced in 1987. Films of WO_3 have been rf reactively sputtered from a W target in an Ar/O_2 atmosphere onto either ultrasonically cleaned microscope slides or ITO coated glass with a sheet resistance ~$130\Omega/\square$. The sputtering system used was a Polaron PT7440 sputter coater with dual mass flow control accessories. A bias voltage up to -200V could be applied to the water cooled substrate. Spectral transmittance values were measured using a Beckman 5240 uv/vis/nir integrating sphere spectrophotometer and solar optical properties determined using a 20 selected ordinate calculation for air mass 2(12).

The electrolyte used to colour and bleach the films was conc. H_2SO_4. An EG and G scanning potentiostat Model 362 was used to apply a constant DC voltage between working and counter electrodes. A Pt electrode was used as the counter electrode and a saturated calomel electrode as the reference.

Figures 6(a) and 6(b) present measured XRD traces in the $2\theta = 20^\circ - 30^\circ$ range for an as deposited film and then following an anneal at 400° C for 24 hours.

This WO_3 film deposited at 150W in an Ar - 10% O_2 atmosphere at 1.0×10^{-3} mbarr for 90 minutes, was ~ 1 μm thick and had a slight blue colouration. It was observed that the film not only changed from being amorphous to crystalline but that this blue colouration disappeared increasing the solar transmittance by 30%. These changes are shown in Figure 7.

When films deposited on the ITO coated glass were heated they went cloudy resulting in an almost total loss of electrochromic sensitivity. This effect has been observed in vacuum evaporated films deposited at pressure $<10^{-5}$ Torr (13). As prepared the films were amorphous and contained ~ 0.5 H_2O molecules per W atom and they exhibited good electrochromic properties in both H+ and Li+ electrolytes. However upon heating to $\sim 300^\circ$ C dehydration and crystallisation occurred destroying electrochromic sensitivity.

For our surfaces it was observed that the most predominant colour changes occurred between 2.0V and 3.0V and that the larger the O_2 content in the film the greater the injected charge indicating an increase in film resistance. On the reverse cycle a greater potential was required to bleach the film than was necessary to colour it.

The depth of colouration of the films was found to be dependent on the length of time that the potential was applied rather than the value of the potential, as can be seen in Figures 8 and 9 respectively for a film deposited in an Ar - 20% O_2 atmosphere. Potentials greater than 3.0V resulted in no further decrease in solar transmittance for colour times of 15s. However an increase in the time for which the potential was applied and hence the amount of injected charge entering the film increased the colouration until a saturation value was reached after which the film would not colour any further.

Open circuit memory of the film was tested by colouring to saturation and leaving it exposed to the atmosphere for 650 hours. The change in solar transmittance over this time period was 24%. As can be seen in Figure 10 the most rapid change occurred over the first 6 hours where solar transmittance increased by 8%. After 650 hours the film was bleached but on colouration a loss in electrochromic sensitivity was observed.

Film dissolution was investigated by immersing a sample in the electrolyte for 64 hours in an ambient atmosphere. Film thickness was found to have decreased by 14 nm giving a dissolution rate of 5.3 nm/day. In water the dissolution rate of a-WO_3 has been estimated as 2.5 μm/day while in a sealed electrolyte of glycerin-H_2SO_4 held at 50°C values of 2-2.5 nm/day have been obtained[2](14). The reaction taking place is given by

$$a - WO_3 + 6h+ \longrightarrow W+^6 \text{ (solution)} + \tfrac{3}{2}O_2 \qquad (2)$$

The dissolution rate for c -WO_3 is considerably lower (15).

During cycling film colouration was taken to maximum or near maximum resulting in a reduction in the number of possible colouration cycles and in the charge per unit area during colouring and bleaching. Although the degradation mechanism is not fully understood it is believed that these observations are an extension of the dissolution process or voltage enhanced corrosion. It has been shown by other workers that after a number of cycles a-WO_3 films have become granular and less adherent to the substrate, possibly due to the action of hydrogen embrittlement. Device degradation can occur by the internal generation of H_2 gas. At potentials above 2.0V gas evolution was observed at the surface of the film. This can arise if the potential across the device is too high or if the film is coloured to H_xWO_3 where x > 0.28. This results in an electrode-electrolyte interface reaction causing dissolution of H_xWO_3 and formation of H_2 gas (15).

Another degradation problem caused by the acidic electrolyte is the chemical etching of the transparent electrode ITO undergoing electrochemical reduction by (15)

$$In_2O_3 \longrightarrow 2In^{3+} + \tfrac{3}{2}O_2 + 3e^- \qquad (3)$$

In order to prevent this a protective layer can be deposited over areas in direct contact with the electrolyte.

CONCLUSION

Although results achieved using WO_3 in electrochromic devices have proved encouraging, problems such as those mentioned and residual colouring must be overcome. There is a need to find a suitable solid electrolyte that is transparent, conductive and easy to prepare. Workers are focusing attention on the possibility of transparent polymer pastes such as Poly-AMPS (Polymerized 2-acrylamido-2 methylpropanesulphonic acid). The electrochromic properties of nickel oxide are currently being investigated and have shown promise for the use of this material in window devices. It is therefore concluded that more time must be spent in developing new materials and improving current devices if switchable window glazings are to be a prospect for the future.

REFERENCES

1. LAMPERT, C.M. "Electrochromic materials and devices for energy efficient windows". Solar Energy Materials 11 (1984).

2. FAUGHNAN, B.W. and CRANDALL, R.S. "Electrochromic displays based on WO_3." Topics in Applied Physics 40 (1980) 181.

3. SVENSSON, J.S.E.M. and GRANQVIST, C.G. "Electrochromic coatings for smart windows". Solar Energy Materials 12 (1985) 391.

4. BENSON, D.K., TRACY, C.E. and RUTH, M.R. "Solid state electrochromic switchable window glazings". SPIE 502 (1984) 46.

5. ASHRIT, P.V., PELLETIER, D.J., GIROUARD, F.E. and TRUONG, V. "Electrolyte thickness dependence of the electrochromic behaviour of a-WO_3 films." J. Appl. Phys. 58 (July 1985) 564.

6. GREEN, M. and KANG, K.S. "Solid state electrochromic cells: the M-β-alumina/WO_3 system". Thin Solid Films 40 (1977) L19.

7. MIYAMURA, M., TOMURA, S., IMAI, A. and INOMATA, S. "Electrochemical studies of lithium nitride solid electrolyte for electrochromic devices". Solid State Ionics 3/4 (1981) 149.

8. TADA, H., BITO, Y., FUJINO, K. and KAWAHARA, H. "Electrochromic windows using a solid polymer electrolyte". Solar Energy Materials 16 (1987) 509.

9. KANEKO, H., MIYAKE, K. and TERAMATO, Y. "Electrochromism of rf reactively sputtered tungsten-oxide films". J. Appl. Phys. 53 (June 1982) 4416.

10. YOSHIMURA, T. "Oscillator strength of small-polaron absorption in WO_3 ($x < = 3$) electrochromic thin films". J. Appl. Phys. 57 (Feb. 1985) 911.

11. GOLDNER, R.B., MENDELSOHN, D.H. et al. "High near-infrared reflectivity modulation with polycrystalline electromic WO_3 films". Appl. Phys. Lett 43 (Dec 1983) 1093.

12. WIEBELT, J.A. and HENDERSON, J.B. "Selected ordinates for solar radiant property evaluation." J. Heat Transfer 101 (1979) 101.

13. ZELLER, H.R. and BEYELER, H.U. "Electrochromism and local order in amorphous WO_3". Appl. Phys. 13 (1977) 231.

14. RANDIN, J.P. "Chemical and electrochemical stability of electrochromic films in liquid electrolytes". J. Electron. Mat. 7 (1978) 47.

15. LAMPERT C.M. "Failure and degradation modes in selected solar materials: a review."

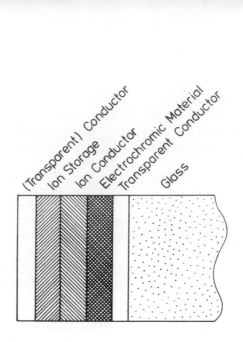

Figure 1 Basic design of an electrochromic device (3).

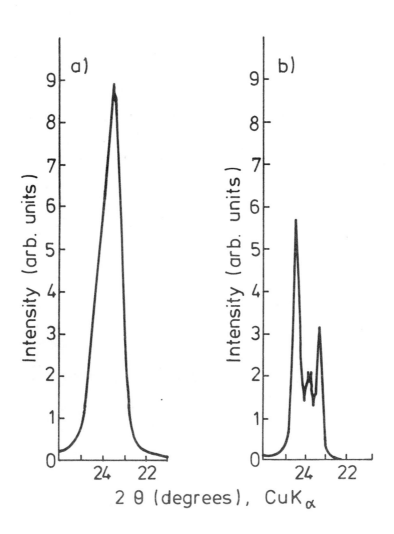

Figure 2 X-ray structure for polycrystalline WO$_3$ films (11).

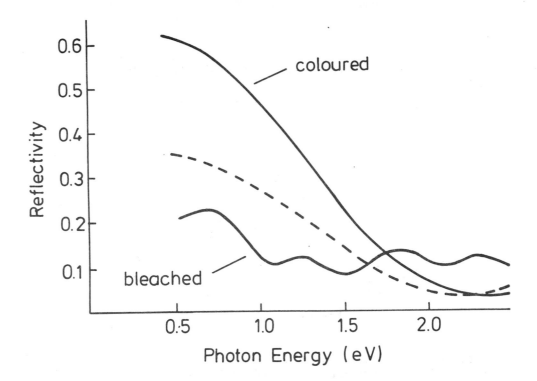

Figure 3   Reflectivity of bleached and coloured
            polycrystalline WO_3 film prepared by rf
            sputtering onto ITO substrate.   Dashed line
            represents thermally evaporated film (11).

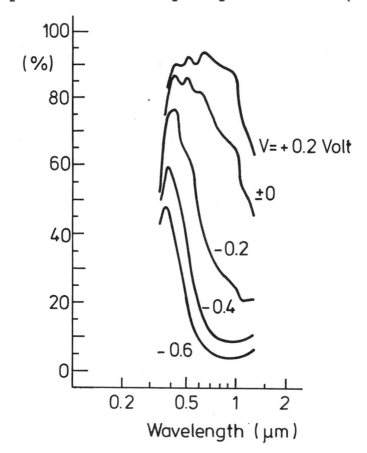

Figure 4   Measured normal spectral transmittance as a function
            of voltage for an electrochromic configuration
            including a liquid electrolyte (3).

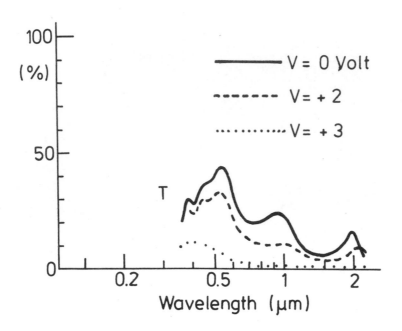

Figure 5 Measured normal spectral transmittance as a function
of voltage for an all-solid-state electrochromic
configuration (3).

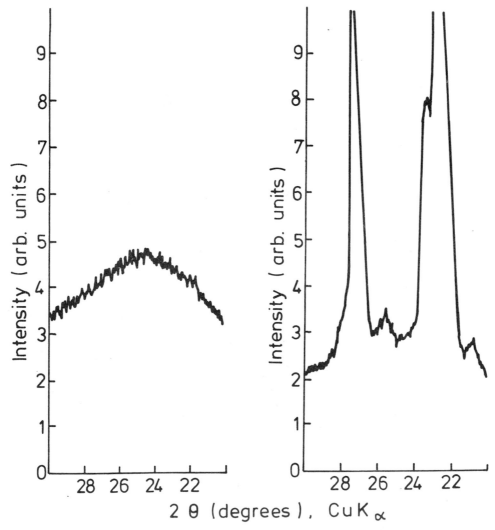

Figure 6 XRD traces of as-deposited and thermally
crystallised WO$_3$ film.

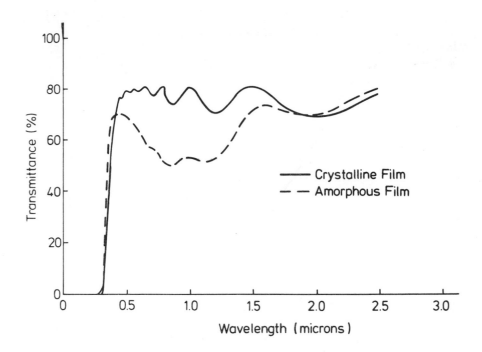

Figure 7   Transmittance as a function of wavelength for an
            as-prepared amorphous WO_3 film subsequently
            crystallised by thermal annealing at $400^\circ C$ for
            24 hours.

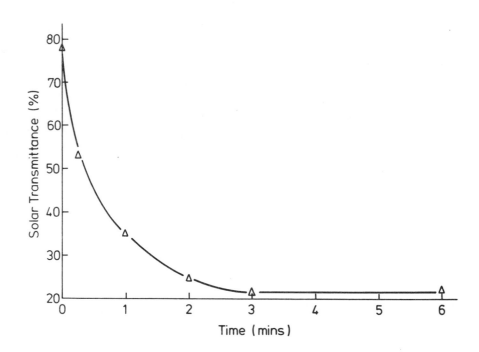

Figure 8   Solar transmittance as a function of time for a
            constant D.C. voltage = 3.OV.

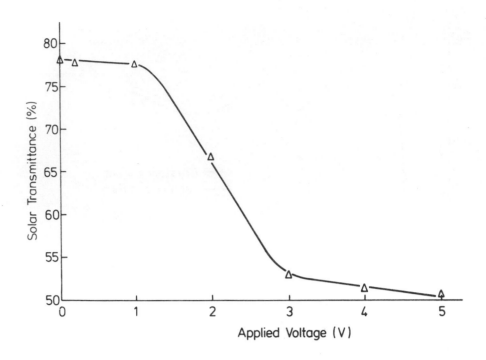

Figure 9   Solar transmittance as a function of applied voltage (colouring time = 15s).

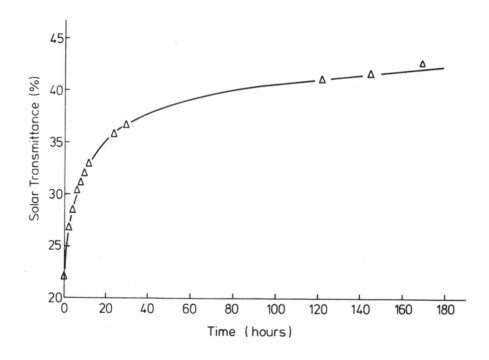

Figure 10   Solar transmittance as a function of time.

NEW FUNCTIONAL WINDOW COATINGS FOR AUTOMOTIVE APPLICATIONS

C.G. Granqvist

Physics Department
Chalmers University of Technology
S-412 96 Gothenburg
Sweden

INTRODUCTION AND MOTIVATION

Coatings can be used to improve the performance of windows[1,2] both in the automotive and the architectural sector. The requirements on these window coatings are similar -- but not identical -- for the two applications. Architectural windows are clearly most central to this Conference. However, a view on the parallel development in the automotive (passenger car) sector may be worthwhile for the following reasons:

- Automobile windows are smaller than typical architectural windows, and the investment required for the production of a specific coating is correspondingly less.

- A person inside a car is always close to a window, and comfort is tied to the performance of the window to a larger extent than in most buildings.

- The expected lifetime is smaller for an automobile window than for a window in a building.

- If a new coating, or an improvement of an earlier used coating, enhances driving safety there is a very strong incentive for employing it.

- The tolerance to increased cost is larger for a window in a (prestige) car than for a window in a (normal) building.

- Conservatism is less of an obstacle in the automotive sector than in the architectural sector, at least in the author's experience.

The above six points all speak in favour of advanced functional window coatings being intro - duced on car windows before they reach the general architectural sector. The fact that automo - bile windows must be curved is not a major problem, and, in fact, coatings can be readily depos- ited onto moderately bent glass by sputter deposition as well as by other technologies. Alterna- tively, coated flat glass can be heated and bent to a desired shape without appreciable deteriora - tion of the radiative properties, which has been demonstrated[3] even for Ag-based thin films.

Below we look at the specific goals one can accomplish by coated automobile windows and discuss the relation to windows in buildings. Separate presentations are given for coatings made with currently available technology and with emerging technology. We then consider recent and as yet unpublished results for a few -- somewhat arbitrarily chosen -- coatings. Specifically, we discuss antireflecting indium-tin-oxyfluoride coatings, electrochromic nickel-oxide-based coatings, and ultraviolet-absorbing zinc-oxide-based coatings.

OPTIONS WITH PRESENT COATING TECHNOLOGY

Present coating technology permits several improvements of the performance of automobile windows. Table I summarizes a number of problems, their principle solutions, and the available coatings. Excessive solar influx, particularly through the windscreen, is a well known problem leading to poor driving comfort. A significant improvement can be reached if the near-infrared solar radiation, carrying about 50 % of the total solar energy, is prevented from entering the car. This must be accomplished without conflict with legal requirements on (normal) luminous transmittance. The best solution is to use a three-layer coating of the type D/M/D, where the metal layer M is Ag (or Cu, Au) and D signifies high-refractive-index dielectric layers (SnO_2, In_2O_3, ZnO, Bi_2O_3, TiO_2, ZnS, etc.). The dielectric layers serve to antireflect the metal so that a high transmittance is induced. The coating is too delicate to be placed on either of the exposed sides of the windows. Placing the coating in proximity to the laminate layer of the windscreen is adequate, though. Another possibility is to work with glass that is absorbing in the near--infrared and, preferably, to prevent the gained heat to be radiated inwards by having a low--emittance coating on the inside of the window. A D/M/D coating is not durable enough, but a layer of SnO_2:F, and presumably also of any of the alternative highly doped oxide semiconductors such as In_2O_3:Sn, has the proper combination of chemical inertness and mechanical durability, at least if the layer is produced by chemical vapour deposition (spray pyrolysis).

TABLE I

Options for coatings on car windows: Present technology

Problem	Solution	Coating*
Excessive solar influx	Reflect at 0.7-3 μm. Absorb at 0.7-3 μm and reflect at 3-50 μm on inside.	D/M/D. In_2O_3:Sn, SnO_2:F,...
Frost	Reflect at 8-13 μm on outside. Electrical heating; high conductance.	SnO_2:F. D/M/D.
Mist	Electrical heating; moderate conductance.	D/M/D, SnO_2:F, In_2O_3:Sn,...
Poor comfort due to thermal gradients	Reflect at 3-50 μm on inside. Electrical heating.	SnO_2:F. D/M/D.

* D/M/D signifies dielectric/noble metal/dielectric.

Frost coverages are causing problems for cars parked outdoors during cold clear nights. The frost is formed preferentially on windows exposed to the sky, since these experience radiative cooling and can drop below the dewpoint faster than surrounding objects.[4] If the surface temperature is below 0°C, a thick and strongly adherent frost layer can form. Radiative cooling is caused by an inbalance between incoming and outgoing radiative powers in the 8-13 μm wavelength range[5] (known as the atmospheric window). It is then obvious that a coating with low emittance in the 8-13 μm interval prevents the window from going below the air temperature, and hence there is no longer any preferential condensation and frostformation. Coatings of SnO_2:F on the outside of the windscreen can yield the desired low emittance[4]. An alternative approach is to apply a coating with high electrical conductivity, which allows rapid heating and deicing of the glass. Suitably protected D/M/D films can have the adequate combination of luminous transmittance and electrical conductivity.[3] Coatings based on heavily doped oxide semiconductors (SnO_2:F, In_2O_3:Sn,...) have too low conductivities to permit acceptable deicing times.

Mist can be removed, or prevented from forming, by electrical heating of the window. The power input can be much lower than for deicing, and both noble-metal based and doped-oxide- semiconductor based coatings can be employed.

Inside the car, the human body is close to the windows -- particularly one of the side windows -- and, if the temperature of the glass is widely different from the body temperature, this will be experienced as discomfort. The situation can be improved by decreasing the radiative exchange between the body and the glass, which is most readily accomplished by a low emittance coating on the inside of the window. SnO_2:F is again a suitable coating. Alternatively, one can use electrical heating of the glass to improve the comfort.

Discomfort due to excessive solar influx and thermal gradients is well known also for windows in buildings. The D/M/D coating, discussed above in the context of windscreens, is equally applicable to windows in buildings in warm or temperate climates. Heated windows are of interest for buildings, since they are able to eliminate cold draughts and permit more un-restricted use of available space. Frost and mist are not problems in properly designed doubly (or multiply) glazed windows.

OPTIONS WITH EMERGING COATING TECHNOLOGY

Active research efforts in universities and industries make it possible to do projections, or at least inspired guesses, of the coatings that may be used on future automobile windows. The presentation is summarized in Table II, organized in analogy with Table I. The first entry regards the earlier discussed coatings, whose performance will be improved. Recent theoretical work gives clear guidelines as to the limiting optical properties both for noble-metal based[6] and doped--oxide-semiconductor based coatings.[7] Effects of microscopic voids can be used to improve the solar transmittance of metal films.[6] For the semiconductors, the limiting properties are governed by the bandgap, the plasma wavelength, and the effect of ionized impurity scattering.[7]

High transmittance is imperative for the windscreen and other windows on a car, and a low re-flectance at the inside of the windscreen is significant for diminishing disturbing reflections from the instrument panel and other parts of the car. Clearly there is a need for antireflection coatings that can be applied effectively on large surfaces and that are compatible with curved glass. Conventional antireflection coatings, such as MgF_2, are of limited use since they cannot be produced by large-area sputter technology. However, recent work[8] on AlO_xF_y and other metal oxyfluorides has shown that very durable coatings, with a refractive index of ~ 1.4 and low ab-sorption, can be prepared by high-rate reactive magnetron sputtering. The coatings can be heated to the softening point of glass without deterioration. Some recent results[9] for indium-tin--oxyfluoride antireflection coatings are reported below.

Future car windows may use thin chemically surface-hardened glass, which allows double glazing. With this design, it is possible to improve the thermal insulation by use of a low--emittance coating either on the inwards-facing side of the outer glass or on the outwards-facing side of the inner glass. Both noble-metal based and doped-oxide-semiconductor based films can be employed. They can be combined with electrical heating for further comfort and for demist-ing. Fillings with gas (Ar, Kr, SF_6...) can enhance the thermal and acoustic insulation.

The earlier mentioned coatings have an inherent limitation in their properties being static, so that they cannot adjust in accordance with variable demands on transmittance. Currently there are intense research efforts aimed at creating practically useful "smart windows" with optical properties that can be regulated between widely separated extrema.[10] In principle one can exploit materials with electrochromism (i.e. whose properties depend on the strength and direction of an applied electric pulse), thermochromism (i.e. whose properties depend on the temperature), and photochromism (i.e. whose properties depend on the irradiation). Electro-chromic coatings offer the greatest flexibility and are considered first. Among the potential

TABLE II

Options for coatings on car widows: Emerging technology

Purpose	Solution	Coating*
Improved performance	Existing.	Existing.
Transmittance	$\lambda/4$-coating.	AlO_xF_y, $M'O_xF_y$,...
Thermal insulation	Double glazing; reflect at 3-50 µm.	D/M/D, SnO_2:F, ZnO:Al,.
Variable transmittance	Electrochromic.	I_xWO_3, I_xNiO,...
	Thermochromic.	VO_2-based.
	Photochromic glass.	(SnO_2:F,...).
Variable reflectance	Electrochromic on reflecting surface.	I_xWO_3, I_xNiO,...
Ultraviolet protection	Absorption at semiconductor bandgap.	ZnO:Al
(Privacy	Liquid crystal.	In_2O_3:Sn,...)

M' signifies a metal; I signifies H, Li,...

applications in cars, we may mention large sunroofs with variable solar energy throughput and visual contact to the ambience. A prototype device of this kind has been demonstrated.[11] Another potential use is in a band at the top of the windscreen; this band is capable of avoiding dazzling by a rising or setting sun.

Electrochromism is known in oxides based on W, V, Ni, Mo, Ir, etc., and in numerous organic substances.[12] The change in the optical properties is caused by the injection or extraction of mobile ions. A material colouring under injection (extraction) is referred to a cathodic (anodic). Absorptance modulation as well as reflectance modulation are possible. A practical window coating should comprise an electrochromic thin film integrated in our all-solid-state multilayer configuration. Figure 1 illustrates a principle design with two outer transparent conductors, required for applying the electric field, an electrochromic layer, an ion conductor, and an ion storage. Coloration and bleaching are accomplished when ions are moved from the ion storage, via the ion conductor, into the electrochromic layer or when the process is run in reverse. The ion storage can be another electrochromic layer, preferably anodic if the base electrochromic layer is cathodic, or vice versa. By use of a purely ionic conductor, one can obtain an open circuit memory, i.e. the electric field has to be applied only when the optical properties are to be altered.

We now consider the different components of the electrochromic-based window coating and first look at the actual *electrochromic* layer(s). Amorphous WO_3 displays absorptance modulation and can yield a variation of the transmittance between wide limits.[10,12] Crystalline WO_3 allows a fair degree of reflectance modulation.[10,13] Electrochromic WO_3 shows best durability when operated in conjunction with aprotic electrolytes, preferably containing Li^+ ions. Fine-crystalline NiO_x is a relatively newly discovered electrochromic material permitting absorptance modulation.[14,15] Its durability seems to be significantly better than for WO_3. Some recent

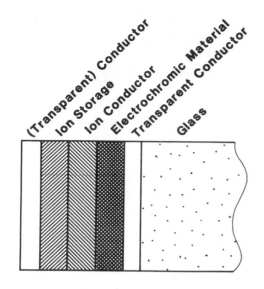

Fig. 1. Principle design of a smart window coating incorporating an electrochromic layer.

work[16] on NiO$_x$-based electrochromic coatings made by reactive dc magnetron sputtering are given later in this paper. The *ion conductor* can be of many kinds; an attractive possibility is to use a polymeric material.[17] Some devices using such ion conductors have been described recently.[18,19] A particularly interesting design, which is currently being studied, is to combine one anodic and one cathodic electrochromic material via a transparent polymeric ion conductor that also serves as a lamination material. The *transparent conductors*, finally, can be of a heavily doped oxide semiconductor. Electrochromic multilayer coatings can be backed by reflecting surfaces so that variable-reflectance devices are accomplished. They are of great interest as rear-view mirrors for cars.[20]

Thermochromic coatings are simple to integrate into a window design. The coating should be capable of diminishing the throughput of radiant energy so that an automatic temperature stabilization is obtained. Thin VO$_2$ films offer interesting possibilities;[21] they can switch from a semiconducting phase below $\tau_c \approx 70°$ C to a metallic phase above this temperature. For utilization of thermochromic VO$_2$ films, it is imperative to find means to decrease τ_c to the vicinity of a normal comfort temperature and to obtain films whose luminous transmittance is acceptably high. It is possible to depress τ_c by replacing V^{4+} by penta- and hexavalent ions, by replacing oxygen by fluorine, by introducing stress through a suitable substrate, and by introducing stress through an overlayer that can serve also to antireflect the VO$_2$ film. Work on SnO$_2$/VO$_2$ tandems has so far yielded $\tau_c \approx 49°$ C and ~ 45 % luminous transmittance.[22]

Photochromic glass becomes absorbing when irradiated by ultraviolet light. Such glass can be used to control the input of radiant energy through a sunroof, and devices of this type have been constructed. In order to be effective, the photochromic glass should be coated with SnO$_2$:F or a similar low-emittance layer on its inwards-facing side. Alternatively, the photochromic glass can be laminated with a normal glass having such a coating.

Ultraviolet radiation is a major reason for the degradation of textiles and polymeric materials used in automobiles. An ultraviolet-absorbing thin film would improve the situation. ZnO:Al coatings capable of combining ultraviolet absorption with infrared reflectance and electrical conductivity are treated below.[23]

Our final example of emerging window technology concerns a foil material[24] whose central part is a polymer with micrometer-sized cavities containing a nematic liquid crystal. This part is sandwiched between polyester foils with electrically conducting In$_2$O$_3$:Sn coatings. The strongly birefringent liquid crystal molecules can be oriented by applying an electric field, and through proper choice of materials one can obtain refractive-index matching for normally incident light so that the material appears transparent. With no field, the molecules are oriented randomly, which leads to index mismatch and strong light scattering. This material is useful for obtaining privacy rather than for efficient control of the throughput of radiant energy.

All of the coatings described here with a focus on automotive applications are useful also for control of energy throughput and glare in buildings.

SOME EXAMPLES OF EMERGING COATINGS

Antireflecting Indium-tin-oxyfluoride coatings (Ref. 9)

In_2O_3:Sn, often referred to as ITO, is one of the most useful and widely employed window coatings of the doped-oxide-semiconductor type. This material, as well as alternative ones of the same type, have a refractive index of ~ 2 in the luminous range, which leads to an undesirably large reflectance. ITO coatings with good properties can be produced by sputtering of an indium-tin-alloy in an $Ar+O_2$ plasma. In our work, such coatings were antireflected by a quarter-wavelength-thick layer of indium-tin-oxyfluoride (ITOF) made by sputtering from the same target in an $Ar + O_2 + CF_4$ plasma. The deposition rate for ITOF was typically 1.2 nms^{-1}. Coatings sputter deposited in a suitably mixed $Ar + O_2 + CF_4$ gas remained stable for at least one year. If the amount of CF_4 was too high or the amount of O_2 was too low, the coatings were hygroscopic and deteriorated in humid air during the course of a few days. These findings are consistent with results for other metal-oxyfluoride coatings.[8] It is conjectured that the basic processes behind the sputter deposition of ITOF are akin to plasma etching of indium-tin with ensuing formation of a non-volatile fluorinated material and simultaneous ejection of carbon in gaseous form.

Spectral transmittance and reflectance were measured for ITOF films, deposited onto different substrates, by use of spectrophotometry, and the complex dielectric function was evaluated from computations based on Fresnel's equations. With an optimized gas mixture, the coatings had a refractive index of 1.4 and little dispersion in the luminous range. The bandgap was ~ 5.9 eV.

Figure 2 shows transmittance and reflectance for a 0.265 μm thick ITO coating with and without a 0.085 μm thick antireflecting ITOF layer. The ITOF/ITO tandem has a luminous transmittance that is as much as 7 % larger than for the bare ITO coating, and the luminous reflectance is correspondingly diminished. A colorimetric analysis showed that the excitation purity, which correlates with the saturation of the colour perceived under ordinary observing conditions, went down by a factor ~ 2 for normal transmission and ~ 4 for normal reflection when the ITOF layer was applied. A similar improvement is not necessary valid for off-normal viewing, though.

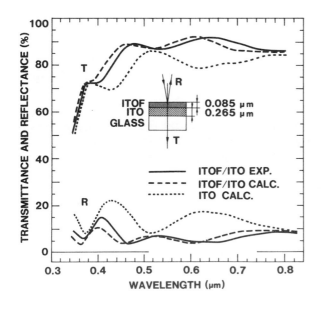

Fig. 2. Spectral normal transmittance and near-normal reflectance as measured for an ITOF/ITO tandem (solid curves), as calculated for the same structure (dashed curves), and as calculated for a bare ITO layer (dotted curves). The calculations were based on the dielectric functions of ITOF and ITO. The sample configuration is shown in the inset.

Electrochromic Nickel-oxide-based coatings (Ref. 16)

Electrochromic NiO_xH_y films were obtained by first making NiO_x through reactive dc magnetron sputtering and then performing electrochemical hydration in KOH. Alternatively, NiO_xH_y could have been made[14] by sputtering in an $O_2 + H_2$ plasma. The deposition rate for NiO_x typically was 0.15 nms^{-1} for coatings that were durable under extended colour-bleach cycling. Transmission electron microscopy yielded that the average crystallite sizes was < 10 nm. Electron diffraction showed patterns consistent with a cubic NiO structure having the same lattice parameter as for the bulk, viz. 0.42 nm, and some fibre texture in the (001) direction.

Colouration and bleaching are expected to proceed according to the reaction

$$Ni(OH)_2 \quad \overset{\text{colour}}{\underset{\text{bleach}}{\rightleftharpoons}} \quad NiOOH + H^+ + e^-$$

i.e. colouration is associated with proton extraction. A practical experiment was conducted by placing a sample, consisting of a 4.25 cm^2 NiO_x layer on an In_2O_3:Sn -coated substrate, in a 1 M KOH electrolyte and applying alternately + 0.1 mA and - 0.1 mA between the sample and a Pt counter electrode. Polarity was reversed each 150th second. After a desired number of cycles, the sample was withdrawn from the electrolyte, rinsed, dried, and subjected to spectrophotometric measurements. When these were completed, the sample could be put back into the electrolyte and run through more cycles.

Figure 3 illustrates the evolution of electrochromism in NiO_x films treated in KOH. The sample comprised layers of $NiO_x H_y$ and In_2O_3:Sn, both being 0.11 μm thick. The dash-dotted curve, referring to an unhydrated NiO_x film, shows that the initial film has low transmittance. The other curves show transmittance of samples in bleached state (zero extracted charge per unit area Q_{ex}) and after heavy colouration with $Q_{ex} \approx 30$ m C cm^{-2}.

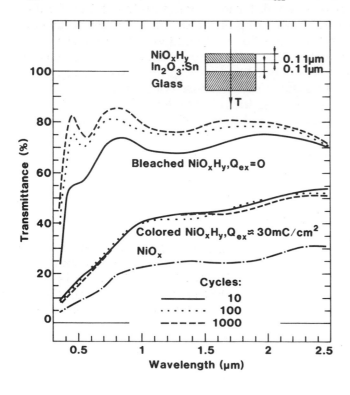

Fig. 3. Spectral normal transmittance for the sample sketched in the inset in bleached and heavily coloured state. Data are shown for different amounts of extracted charge per unit area and for different numbers of colour-bleach cycles.

Solid, dotted and dashed curves signify samples run through 10, 100, and 1000 colour-bleach cycles, respectively. It appears that it takes >> 10 cycles to produce a fully bleached sample. The coloured state, on the other hand, is independent of the number of cycles.

Optical data of the kind shown in Fig. 3, as well as for samples at intermediate colouration, yielded that the luminous transmittance could be modulated continuously and reversibly between 80 and 20 %. The corresponding values for the solar transmittance were 74 and 29 %. The transmittance in the coloured state seems to be dependent on the method for making the NiO_xH_y coating, and the present technique with dc magnetron sputtering in O_2 did not lead to quite as low magnitudes as the earlier employed[14] rf magnetron sputtering in O_2 + H_2.

Ultraviolet-absorbing Zinc-oxide-based coatings (Ref. 23).

ZnO is a semiconductor with a bandgap equal to 3.4 eV. The corresponding values for In_2O_3, SnO_2, and CdO, are 3.75, 3.7, and 2.3 eV, respectively, and hence only ZnO is capable of absorbing a substantial part of the ultraviolet radiation coming from the sun without also absorbing luminous radiation. Further, ZnO can be doped[25] so that a low thermal emittance and a high electrical conductivity are achieved, and hence ZnO-based thin films are alternatives to other coatings of the heavily-doped-oxide-semiconductor type. ZnO-based coatings are of particular interest since they consist of cheap and abundant elements, are non-toxic, and are readily produced by sputtering technology.

ZnO:Al coatings were made on CaF_2 substrates by simultaneous rf magnetron sputtering of ZnO and dc magnetron sputtering of Al. The sputter ambient was weakly reactive with an Ar/O_2 ratio larger than 400. The substrate was mounted perpendicular to the plane of the targets and was rotated at a speed that ensured mixing on an atomic level. Quantitative elemental analysis was obtained by Rutherford Backscattering Spectrometry.

Spectral transmittance and reflectance were measured in the 2-5 eV range for ZnO:Al coatings containing up to ~ 2 at.% Al. These data were used to evaluate the complex dielectric function, from which the absorption coefficient was derived. Figure 4 shows results for samples with different degrees of doping. The bandgap goes from ~ 3.4 eV for low doping up to ~ 3.9 eV for the highest doping. This widening is accompanied by a smearing of the transition between low and high absorption. It is well known[26] that the bandgap shift due to doping is the net effect of a widening caused by electrons occupying the lowest states of the conduction band and a narrowing caused by manybody effects on the conduction and valence bands. A quantitative theory of the bandgap shift has been formulated. It incorporates the effective masses of both the conduction and valence bands of ZnO. The polar character of ZnO was included by employing recent theoretical work. [27] This theory could provide a fully quantitative model for the bandgap shift inherent in Fig. 4 without the need to invoke any free parameter. Practical ultraviolet-absorbing and infrared-reflecting window coatings may comprise ZnO/ZnO:Al tandems.

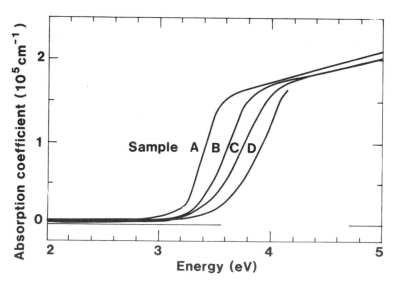

Fig.4. Spectral absorption coefficient for ZnO:Al coatings with different Al contents. Samples A-D contain 0, 0.95, 1.40 and 2.14 at.% Al, respectively.

REFERENCES

1. C.G. Granqvist, in Physics of Non-conventional Energy Sources and Materials Science for Energy, edited by G. Furlan, N.A. Mancini, A.A.M. Sayigh and B.O. Seraphin (World Scientific, Singapore, 1987), pp. 217-302.

2. C.G. Granqvist, in Physics and Technology of Solar Energy, edited by H.P. Garg et al. (Reidel, 1987), Vol. 2, pp. 191-276.

3. J. Szczyrbowski, A. Dietrich and K. Hartig, Proc. Soc. Photo-Opt. Instrum. Engr. 823, 21 (1987).

4. I. Hamberg, J.S.E.M. Svensson, T.S. Eriksson, C.G. Granqvist, P. Arrenius and F. Norin, Appl. Opt. 26, 2131 (1987).

5. T.S. Eriksson and C.G. Granqvist, Appl. Opt. 21, 4381 (1982).

6. G.B. Smith, G.A. Niklasson, J.S.E.M. Svensson and C.G. Granqvist, J. Appl. Phys. 59, 571 (1986).

7. I. Hamberg and C.G. Granqvist, J. Appl. Phys. 60, R123 (1986).

8. G.L. Harding, Solar Energy Mater. 12, 169 (1985); Thin Solid Films 138, 279 (1986).

9. S.-J. Jiang, Z.-C. Jin and C.G. Granqvist, to be published.

10. J.S.E.M. Svensson and C.G. Granqvist, Solar Energy Mater. 12, 391 (1985).

11. Popular Science., Dec. 1987, p. 70.

12. C.M. Lampert, Solar Energy Mater. 11, 1 (1984).

13. R.B. Goldner, R.L. Chapman, G. Foley, E.L. Goldner, T. Haas, P. Norton, G. Seward and K.K. Wong, Solar Energy Mater. 14, 195 (1986).

14. J.S.E.M. Svensson and C.G. Granqvist, Appl. Opt. 26, 1554 (1987).

15. M.K. Carpenter, R.S. Conell and D.A. Corrigan, Solar Energy Mater. 16, 333 (1987).

16. W. Estrada, A.M. Andersson and C.G. Granqvist, to be published.

17. M.B. Armand, Ann. Rev. Mater. Sci. 16, 245 (1986).

18. J.R. Stevens, J.S.E.M. Svensson, C.G. Granqvist and R. Spindler, Appl. Opt. 26, 3489 (1987).

19. H. Tada, Y. Bito, K. Fujino and H. Kawahara, Solar Energy Mater. 16, 509 (1987).

20. F.G.K. Baucke, B. Metz and J. Zauner, Phys. unserer Zeit 18, 21 (1987).

21. S.M. Babulanam, T.S. Eriksson, G.A. Niklasson and C.G. Granqvist, Solar Energy Mater. 16, 347 (1987).

22. S.M. Babulanam, W. Estrada, M.O. Hakim, S. Yatsuya, A.M. Andersson, J.R. Stevens, J.S.E.M. Svensson and C.G. Granqvist, Proc. Soc. Photo-Opt. Instrum. Engr. 823, 64 (1987).

23. B.E. Sernelius, K.-F. Berggren, Z.-C. Jin, I. Hamberg and C.G. Granqvist, Phys. Rev. B, to be published.

24. P. van Konynenburg, S. Marsland and J. McCoy, Proc. Soc. Photo-Opt.Instrum. Engr. <u>823</u>, 143 (1987).

25. Z.-C. Jin, I. Hamberg and C.G. Granqvist, Appl. Phys. Lett. <u>51</u>, 149 (1987).

26. I. Hamberg, C.G. Granqvist, K.-F. Berggren, B.E. Sernelius and L. Engström, Phys. Rev. B<u>30</u>, 3240 (1984).

27. B.E. Sernelius, Phys. Rev. B<u>34</u>, 5610 (1986).

THE PHYSICAL PROPERTIES OF PYROLYTICALLY SPRAYED TINDIOXIDE COATINGS

H. Haitjema

Delft University of Technology
Applied Physics Department
P.O. Box 5046
Delft, The Netherlands

INTRODUCTION

Tindioxide layers are electrically conducting and infrared reflecting coatings with a wide range of applications. In the field of solar energy conversion they can be used as a conducting anti-reflection electrode for photovoltaic cells or as a spectrally selective layer for photothermal conversion.

Tindioxide acts as a spectral window. Light is transmitted in the wavelength region confined by λ_{gap} and λ_{plasma}. Below λ_g band gap absorption occurs, above λ_p reflection due to free electron plasma behaviour takes place. As most of the solar spectrum is in the 0.3 - 2.0 µm region and thermal radiation is in the infrared above 2 µm λ_p should preferably have a value of $\sim$ 2 µm. This can be achieved by suitably doping the tindioxide with antimony or fluorine, which enhances the metal character of the layer.
Tindioxide layers are mechanically and chemically highly stable, also at high temperatures, and can be produced in a relatively cheap and simple way by spray pyrolysis [1-4].

The major problem when producing tinoxide layers is a relatively high emissivity, which is related to a limited electrical conduction in the layer. In this paper the cause of this limited conductivity is more thoroughly studied and the results of optimizing the spectral-selective properties will be presented.

PREPARATION

The coatings are produced by a spray pyrolysis method as described in an earlier paper [1]. We used three different spray solutions:
  1. a solution of $SnCl_4$ in water and alcohol with NH_4F as a dopant
  2. a solution of $(C_4H_9)_2 SnO$ in acetic acid with $CF_3 COO H$ as a dopant
  3. a solution of $(C_4H_9)_3 SnF_3 (C_2H_3O_2)_2$
To obtain layers of 0.6 µm thickness or more on window glas or pyrex the substrate is heated and sprayed several times to minimise a large temperature fall during spraying. If spraying is carried out on black enamelled steel the substrate can be heated to a higher temperature and the coating can be deposited in one go.

MEASURING TECHNIQUES

Reflectance and transmittance in the 0.3 - 2.5 µm region are measured with a Perkin-Elmer Lambda 9 Spectrophotometer, which is equipped with an integrating sphere. The reflectance in the infrared region (2.5 - 50 µm) is measured with a Perkin-Elmer 883 IR-Spectrophotometer with a specular reflectance attachment. The quantities important for solar energy application, thermal emittance ε and solar absorptance α, are measured separately with apparatus as described by van der Ley [5].

The electrical properties, resistivity and Hall coefficient, are measured according to the van der Pauw method (van der Pauw, 1958) using circular-shaped samples in a 0.6 T magnetic field. A sample holder is used which enables the sample temperature to be varied between 77 and 600 K in vacuum.

Layer thickness is usually calculated from the optical extrema using a method described in an earlier paper [1]. When the layer thickness is less than about 100 μm a Tencor alphastep 200 stylus apparatus is used. In that case a step is made by etching the coating with Zn powder and HCl.

Determination of optical constants

The determination of optical constants as a function of wavelength is essential when comparing the optical and electrical properties. In the 0.3 - 2.5 μm wavelength region the complex refractive index is determined from reflection and transmission measurements using the R-T method [6].

In the 2.5 - 25 μm region, where the substrate is absorbing, we use an iterative method to determine the optical constants from the reflection spectrum only.

A first approximation is obtained by applying a Kramers-Kronig relation in the 2.5 - 15 μm region:

$$\phi\,(\omega_o) = \frac{\omega}{\pi} \int_o^\infty \frac{\ln R(\omega) \,-\, \ln R(\omega_o)}{\omega^2 \,-\, \omega_o^2}\, d\omega \qquad (1)$$

The spectrum is extended to 0.3 μm by calculating R for a semi-infinite medium from the refractive index obtained by the R-T method using:

$$R = \frac{(n - 1)^2 + k^2}{(n + 1)^2 + k^2} \qquad (2)$$

Above 25 μm the spectrum is extrapolated using the Hagen-Rubens relation

$$R = 1 - (\frac{2\omega\rho(DC)}{\pi})^{1/2} \qquad (3)$$

When the reflection surface is semi-infinite the complex refractive index is given by:

$$n(\omega) = n - ik = \frac{1 - R - 2i\,\sqrt{R}\,\sin\phi}{1 + R - 2\,\sqrt{R}\,\cos\phi} \qquad (4)$$

Now (4) is used as a first approximation. Using n and k from (4) together with the film thickness and the substrate optical, constants we can calculate the reflectance from the film-substrate system R (ω). When this calculated reflectance differs from the measured spectrum R(ω) we can calculate a 'synthetic' spectrum R"(ω) which can be expected to give the correct optical constants when using (1) and (4):

$$R''(\omega) = R(\omega)\ .\ (\frac{R\ (\omega)}{R'\ (\omega)}) \qquad (5)$$

This procedure is repeated until a spectrum $R''(\omega)$ is obtained which gives optical constants by ((1) and (4)) which give the measured reflectance $R(\omega)$ when the reflectance of the film-substrate system is calculated. From the optical constants the complex resistivity is $\rho(\omega)$ is deduced using the definition:

$$\varepsilon_r(\omega) = (n - ik)^2 = \varepsilon_{r\infty} - \frac{1}{\rho(\omega)\varepsilon_o\omega} \qquad (6)$$

Where $\varepsilon_{r\infty}$ is the relative permittivity at $\omega=\infty$, which is about 4.1 for SnO_2.

EFFECT OF DOPING CONCENTRATION

A series of layers with increasing doping concentration has been obtained using solution 1. The results of electrical measurements at room temperature on these layers are depicted in figure 1. Figure 1 shows that both the electron density n_- and the mobility μ increase with temperature.

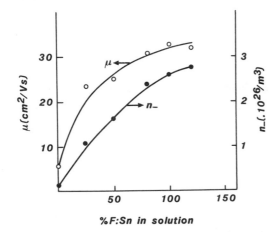

Figure 1   Electrical mobility and electron density

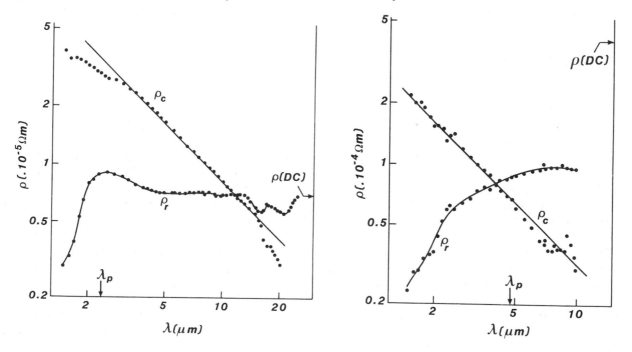

Figure 2   Real and imaginary part of resistivity

The real and complex part of the complex resistance, as defined in (6) are given for the undoped and the 100 % F-doped case in figure 2.

From results as given in Figs. 1 and 2 we can derive some useful quantities using the Drude theory, which can be written as:

$$\rho(\omega) = \frac{\gamma + i\,\omega}{\varepsilon_o\,\varepsilon_{r\infty}\,\omega_p^2} \qquad (7)$$

Here ω_p is the plasma frequency which can be defined as:

$$\omega_p^2 = \frac{n_-\,e^2}{\varepsilon_o\,\varepsilon_{r\infty}\,m_{eff}} \qquad (8)$$

with e being the electron charge and m_{eff} the effective electron mass. γ is the relaxation frequency given by

$$\gamma = \frac{e}{m_{eff}\,\mu} \qquad (9)$$

Comparison between the measured electron density and the complex part of $\rho(\omega)$ gives a value for m_{eff}. The real part of $\rho(\omega)$ can be directly compared to the DC-resistivity

$$\rho(DC) = (n_-\,e\,\mu)^{-1} \qquad (10)$$

In fig. 2 it is shown that for the doped sample the $\rho(DC)$ value is in accordance with the AC-value, while for the undoped sample the AC-value is much lower.

We can account for this difference by defining an optical mobility by:

$$\mu(opt) = (n_-\,.\,e\,.\,\rho_R\,(\lambda=5\mu m)) \qquad (11)$$

The results for the samples depicted in Fig. 1 are summarised in table 1.

Two noteworthy results can be seen in table 1: The effective mass depends on the electron concentration and the optical mobility is higher than the DC-electrical mobility when the electron density is low. The first result indicates that the conduction band will not be parabolic. The difference between AC and DC-mobility can be explained when we consider the effect of grain-boundary scattering, which will affect the DC-properties rather than the optical properties, which will be mainly determined by the bulk.

As the electron energy's are distributed according to degenerate Fermi-Dirac statistics the Fermi-energy will be dependent on the electron density. Now the difference between optical and DC-mobility can be explained as follows: when the electron density is low the electrons will have less energy than the grain-boundary gap energy and the conduction will be determined by the tunnelling of electrons through the grain-boundarys. When the electron density is increased the Fermi-level will raise above the boundary gap energy and the conduction will be determined by the properties of the bulk crystalles.

TABLE 1.   Properties of coatings with different F-doping

%F:Sn in solution	%F:Sn in coating	n_{-} ($.10^{26}/m^3$)	μ_{el} ($.cm^2/Vs$)	λ_p (μm)	m_{eff} ($.m_{el.}$)	μ_{opt}
0	0	0.293	5.6	4.80	0.15	$\sim$ 23
25		1.21	23.4	3.76	0.26	24.8
50	0.65	1.78	25.1	2.70	0.28	25.7
80	0.74	2.46	31.2	2.29	0.28	31.3
100	0.96	2.71	33.0	2.38	0.34	32.8
130	1.04	2.77	32.9	2.26	0.31	32.0

TEMPERATURE-DEPENDENCE OF MOBILITY

The mobility of three samples having low, average and high mobility respectively has been measured as a function of temperature. The results are shown in Fig. 3. In this figure we see that when the mobility is low it increases with

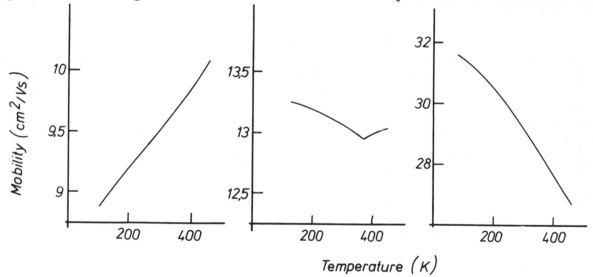

Figure 3   Temperature dependence of mobility

temperature, while when the mobility is high it decreases with temperature. This different behaviour can be explained if we consider the various scattering mechanisms that are operative to determine the measured mobility. The effective mobility μ_F is given by [7,8]

$$\frac{1}{\mu_F} = \frac{1}{\mu_c} + \frac{1}{\mu_l} + \frac{1}{\mu_{gb}}$$

(12)

where μ_c takes into account temperature-independent scattering of neutral and ionised impurities, μ_l is due to lattice scattering $\mu_l \sim T$ and μ_{gb} accounts for grain boundary scattering and can be written as $\mu_{gb} = \mu_0 \exp(-\phi/kT)$ where ϕ is

73.

the inter grain barrier height. μ_0 is a prefactor which is proportional to the electron velocity and so to the electron density.

Now the different behaviour in Fig. 3 can be explained. When the mobility is high, the grain boundaries have little influence so the mobility is mainly determined by impurity scattering and lattice scattering, resulting in a decrease of the mobility with temperature. An increase of μ_F with temperature, which occurs when the mobility is low, can only be explained by grain boundary scattering. In this case μ_1 and μ_c are less important and will have about the same values as in the case of high mobility. So if a high electrical mobility is desired, first the influence of grain boundaries has to be minimised. Further improvement might be achieved by minimising the number of neutral impurities in the the layer.

RESULTS OF OPTIMIZING

We optimised the Fluorine-doping (to obtain $n_- = 3.5 \cdot 10^{26}/m^2$) and also the substrate temperature during spraying (to make μ as high as possible). Results are obtained as shown in table 2. The substrate temperature must be high because during spraying the substrate temperature may drop up to hundred ^oC. On the other hand window glass cannot be heated above 600 ^oC because weakening occures at higher temperatures.

TABLE 2.   Results of optimizing

Substrate	spray solution	T_s (^oC)	ε_h	α	d (μm)	n_- ($.10^{26}/m^3$)	μ (cm^2/V_s)
Glass	1	600	0.19	0.87	0.50	3.2	35
Pyrex	2	640	0.17	0.88	0.62	2.7	42
Pyrex	3	600	0.15	0.87	0.60	3.5	45
Black Enamel	2	740	0.15	0.91	0.68	3.5	45

T_s is substrate temperature, ε_h is total hemispherical emissivity, α is the solar absorption coefficient and d is layer thickness.

Solution 1: $SnCl_4$ in water and alcohol with NH_4F

2: $(C_4H_9)_2$ SnO in acetic acid with CF_3COOH

3: $(C_4H_9)_3$ SnF_3 $(C_2H_3O_2)_2$ in alcohol

Best results are obtained with the organic solutions. The best coatings have a specific resistance of about $4 \cdot 10^{-6}$ Ωm, which is competitive to In_2O_3-coatings.

ACKNOWLEDGEMENTS

The authors wish to thank the TPD (Institute of Applied Physics TNO-TU) for the use of their spectrophotometers. This investigation in the program of the Foundation for Fundamental Research on Matter (FOM) has been supported (in part) by the Netherlands Technology Foundation (STW).

REFERENCES

[1] Haitjema, H. and J. Elich, (1987). The physical properties of fluorine-doped tindioxide films and the influence of ageing and impurity effects. Sol. En. Mat., 16, 79-90.

[2] Simonis, F., A.J. Faber and C.J. Hoogendoorn, (1987). Porcelain Enamelled Absorbers, coated by Spectral Selective Tin oxide. J. Sol. Eng., 109, 22-25.

[3] Simonis, F., M. van der Ley and C.J. Hoogendoorn,(1978). Physics of doped tindioxide films for spectral-selective surfaces. Sol. En. Mat., 1, 221-232.

[4] Karlsson, T., A. Roos and C.G. Ribbing, (1985). Influence of spray conditions and dopants on highly conducting tin dioxide films. Sol. En. Mat., 11, 469-478.

[5] Ley, M. van der, (1979). Spectral-selective surfaces for the conversion of Solar Energy, Ph.D. dissertation, Delft University Press, Delft.

[6] P.O. Nilsson, (1968). Determination of optical constants from intensity measurenments at normal incidence. Appl. Opt., 7, 435-441.

[7] Kohle, S., S.K. Kulkarni, M.G. Takwale and V.G. Bhide, (1986). The electrical conduction in spray CdS films. Sol. En. Mat., 13, 203-211.

[8] Stapinski, T., E. Leja and T. Pisarkiewiez, (1984). Point defects and their influence on electrical properties of reactive sputtered Cd_2SnO_4 thin films. J. Phys.D: Appl. Phys., 17, 407-413.

SELECTIVE SOLAR ABSORBER SURFACES - PRESENT AND FUTURE

Dr. J.J. Mason

Inco Selective Surfaces
Wiggin Street
BIRMINGHAM
B16 0AJ

INTRODUCTION

In the 30 years since the introduction on a commercial scale of selective solar absorbing surfaces, several types have been established in the market place. All commercially available surfaces in quantity production are directed to the only significant application area, that of water heating for domestic applications. This paper reviews the characteristics of those surfaces commercially available and highlights the importance of product form in establishing a market position. Although no new products have been commercialised in the last 5 years, some possible candidates for future production are discussed.

ABSORBER COATINGS: GENERAL CONSIDERATIONS

In commercial terms an ideal selective absorber has a value of solar absorptance greater than that of competitive non-selective black paints or non-selective surfaces finishes, i.e. greater than 0.95 combined with a thermal emittance of less than 0.15. Often the ratio is taken as a measure of selectivity and, whilst this may be so, it does not necessarily reflect the utility of a surface. Taking typical properties of 0.95 and ε of 0.10, a 2% increase in both parameters will give the same α/ε ratio but almost a 2% increase on solar collector output. Therefore any increase in absorptance outweighs a corresponding increase in emittance as long as both parameters stay within the criteria outlined initially.

Clearly the cost of the product must be competitive. Since most coatings are relatively thin, the intrinsic material component of the cost is generally low, other components including capital equipment cost, production cost and development costs being the major items. Typically a selective absorber surface must be capable of being produced at a cost of not more than £2.00/m2 and preferably less than £1.00/m2. In order to offset the high capital cost of equipment and low initial solar sales many products have used equipment which had doubled for the production of black or coloured surfaces for decorative applications.

Many of the surfaces are produced by batch processes which often mean high production costs and poor reproducibility of properties However, surfaces produced by continous processes can often lead to greater uniformity of product combined with reduced labour costs.

Some processes can be used in batch and continuous product form.
In the first example it is a finished process applied to a
fabricated solar heat exchanger panel. Because of the fragility
of absorber surfaces and the difficulties and costs of packa-
ging such services are usually provided locally, available
within a region to several solar collector manufacturers. In
the second case a facility is set up to continuously coat strip
or coil which is sold more widely as a semi-finished product
for forming into fins or heat exchanger plates by the collector
manufacturer. The surface property requirements for the two
routes differ. In the batch process the surface merely has to
withstand the collector service conditions, typically a maximum temperature for a single glazed collector of 200°C. The
coil product has to withstand the fin fabrication process often
involving pressing, roll forming and brazing operation which if
superheat is not controlled can give temperatures well in
excess of 300°C.

A wide range of materials is used as the absorber in flat plate
collectors including copper, aluminium, stainless steel and
mild steel. Selective coatings should be capable of application
to as many substrates as possible. However, many coatings,
because of the production route and the nature of the coating
are only applicable to one substrate. Two types of coating can
be used on any substrate. Thin selective absorbing foils, often
with adhesive backing can be applied to flat surfaces, whilst
paint is even more widely applicable. Unfortunately, from an
industry viewpoint, no usefully selective paint, having thickness insensitive properties, has yet to be developed. The lack
of a durable infrared transparent binder has limited and will
continue to prevent the development of an useful solar selective
paint.

COMMERCIALLY AVAILABLE COATINGS

Commercially produced coatings have been catergorized in terms
of their production route. The two main processes are electrodeposition and chemical conversion. In electroplating, metals
or co-deposition of each are plated onto the absorber surface or
dielectrics into a porous oxide coating on the absorber.
Conversion coatings rely on the formation of a surface oxide by
conversion of the substrate itself. Under oxidizing conditions
it occurs over a narrow range, called the transpassive region
between the chemical dissolution of the substrate and the pass-
ivation of the surface. Just a few of the processes are listed
in Table One.

TABLE ONE

Various black conversion coatings for metals.

Metal substrates	Process	Conversion coatings
Al	Acidic copper nitrate	Oxide
	Permanganate	Complex molybdate
	Hot molybdate-chloride	
Steel	Hot alkaline oxidation	Oxide
	Phosphating	Fe-zinc phosphate
	Molten Sodium dichromate	Oxide
Stainless Steel	Hot alkaline oxidation	Oxide
	Molten sodium dichromate	Oxide
	Hot acid oxidation	Oxide
Cu	Hot alkaline oxidation	Oxide
	Selenium black	Selenide
	Ammonium sulphate	Sulphate

Electrodeposited Coatings

Black Chrome

Starting from the observation by workers at NASA of the useful
selective absorbing properties produced with propriety black
chrome plating baths, black chrome has established itself as the
market leader in selective surfaces in the U.S.A. and is also
widely produced and used in other countries. The process was
originally developed commercially as a decorative coating and
this has clearly helped in its development for solar applications.
Black chrome plating baths consist of a chromic acetic mixture
with additives of trivalent chrome and iron. The coating is
deposited at high current densities, about 2000A/m2 and low
bath temperatures (20-30°C). Special shielding techniques have
to be used to produce a uniform coating thickness over a solar
collector. Typical properties of conventional coatings are an
absorptivity of 0.96 combined with a termal emittance of 0.12.
The coating is composed of a top layer of loosely packed particles of Cr_2O_3. The size of this top structure is of the order
of solar wavelengths, thus allowing multiple reflections which
enhance the absorption process. The rest of the film is metal-
dielectric composite consisting of closely packed chromium particles separated by voids and each probably covered with a shell
of Cr_2O_3. These films have adequate termal stability for flat
plate collectors but heating at 350°C or more does cause oxygen
to diffuse into the film, causing a slight reduction in absorptance. Black chrome can be applied to a wide range of metallic
substrates. Because of the porous nature and thinness of the
coating, it is normally recommended that an undercoat of nickel
is provided to give adequate durability. Tests have shown that
a semibright or dull nickel undercoat is preferred. The generally recommended minimum undercoat thicknesses of nickel are
12.5 micron for steel and aluminium absorbers and 5 micron for
copper. However, the most popular substrates for black chrome
is copper.

It is applied to finished solar absorbers by certain decorative platers in Europe and U.S.A. Black chrome plated copper strip in widths up to 1m and thicknesses down to 50 micron is also produced in North America and recently a facility has been commissioned in India. The stabilities of black chrome is such that it can be soldered, using high temperature solders, without serious degradation.

Coloured Anodized Aluminium

Selective surfaces on anodized aluminium have developed from coloured anodized aluminium for decorative and architectural applications. As anodized, the oxide film on aluminium is porous and receptive to colouring by several techniques. Organic dyes and inorganic pigments have been used to fill the pores partly or totally, thus directly colouring the surface. The range of colours available and their stability limit the use of this method of colouring. Alternatively colours can be developed by plating metallic particles within the oxide pores from acidic metal salt solutions. Most frequently the sulphate is used whilst metals, including nickel, cobalt, copper and tin, have all been examined and found to be suitable. Most of these finishes are particularly light fast. This excellent stability is partly due to the metallic nature of the pigment and partly that, under the influence of the a.c. plating conditions, the pigmentation occurs at the base of the pores. For this reason relatively thin anodized coatings of about 5 micron can be coloured to the same shades as thicker films. Of course such anodized layers (5 micron thick) have a relatively high thermal emittance and it is necessary to use much thinner anodized layers for the first stage in the production of a selective absorber.

The normal production technique, which has been described by amny workers consists of anodizing clean aluminium in a weak phosphoric acid solution and then, after washing, plating under a.c. conditions in a buffered nickel sulphate solution. The surfaces are then rinsed and sealed by immersion in boiling water. Properties claimed by manufacturers in Sweden, Japan and Germany are typically α -0.95 with ε - 0.15.

Although this surface treatment is obviously restricted to aluminium substrates, much ingenuity has been used to widen its marketability. The treatment can be applied to all aluminium solar absorber plates and, whilst water side corrosion can be minimized in mixed metal systems by the careful use of inhibitors, this feature of aluminium does prevent its use in several countries. One producer of the selective surface has developed an absorber fin consisting of two sheets of aluminium along the centre of which is a copper tube. The composite is rolled to form a fin in which copper is metallurgically bonded to the aluminium and completely surrounded by aluminium. As the inner surface of the copper has a layer of oxide and oil it does not bond together and can be reinflated after the rolling process. The fin can subsequently be anodized and blackened to give good selective properties.

The fin is sold widely to solar collector manufacturers for on
site fabrication. The anodizing process has also been applied
to aluminium sheet, up to 1.2m wide and 0.2mm thick, for use
in air collectors and passive solar systems.

Chemical Conversion Coatings

Coloured Stainless Steel

The excellent durability of stainless steels, to both air and
water side corrosion, makes them important candidates as
absorber plate materials for solar collectors. This advantage
is complemented by the relative ease by which selective absorbing surfaces can be developed on stainless steel. Oxide films
can be grown chemically on stainless steels using a hot aqueous
chromic-sulphuric acid solution. As formed these films are
relatively soft and porous. Electron micrographs have shown the
pore fraction to be up to 30%. The commercialization of the
process, initially for decorative and architechural applications
rested on the development of a process for hardening the films
therefore rendering them resistant to handling. The hardening
process involving the electrodeposition of chromic oxide within the pores, has little effect on the optical properties of
the surface film. Typical properties of the coating are an
absorptance of 0.90 combined with an emittance of 0.12.

Because of the poor thermal conductivity of stainless steel,
there are relatively few economical designs open to it when used
in flat plate solar absorbers. The main design is an inflated
heat exchanger consisting of two thin sheets of stainless steel
seam and spot welded to contain water. This allows the maximum area of water contact with the absorbing surface. Panels
can be fabricated from coated sheet or alternatively welded
panels can be surface converted. These panels are popular in
Japan but have limited appeal in Europe and North America.

Nickel Oxide

An oxide conversion process has been developed for producing a
selective surface absorbing surface on pure nickel. The thin
oxide, typically 0.3 micron thick, is produced on a micro
roughened nickel substrate. Typical properties for the coating
are an absorptivity of 0.95 - 0.98 with an emittance range from
0.08 to 0.12. Because of the cost of nickel it cannot, of
course, be considered as an absorber material itself but, as
described earlier, nickel is often used as a protective under-
layer for other selective coatings. The nickel oxide conversion coating can therefore be used to treat nickel-plated
components in a batch process. However, advantage is taken
of the ability of the process to be used in a continuous mode
to surface treat nickel foil 12 micron thick and nickel-plated
copper strip 0.1 and 0.2mm thick. The nickel foil, available
in widths up to 0.5m, can be coated with adhesive and fixed to
any metal absorber plate provided that it has a suitable surface
profile. The adhesive isolates both chemically and electrically
the underlying base material, thereby minimizing potential

corrosion problems. Structual adhesives have been used to bond the foil to Al or Cu strip which can then be processed to make a fin.

FUTURE SELECTIVE ABSORBERS

With the predicted increase in the use of holograms in large volume applications such as advertising, printing and daylight control, it is likely that processes for production and replication of holograms will become relatively cheap. Furthermore the availability of spare capacity on equipment for hologram production will stimulate the development and commercialization of new products.

Workers at NPL have drawn attention to the production of so-called Moth-Eye structures which on non conductors can have antireflective characteristics and on metal surfaces can be spectrally selective. Moth-Eye structures consist of a surface with a regular geometric array of sub micron protuberences. They are produced in a photoresist by exposure to a straight line interference fringe pattern from two coherant laser beams. The photoresist is exposed twice to the radiation, the photographic plate being rotated 90° in its plane between exposures. In order to have a useful product it is necessary to replicate the structure onto an alternate surface. This can be done by metallizing the photoresist. Unfortunately this destructs the photoresist but second generation replicas can be produced from the original metallic master. These replicas can be used as embossing plates to reproduce the structure in thin plastic film. Upon metallizing the plastic film is transformed into a selective absorber. NPL workers have reported solar absorptances of 96% and thermal emittance of 0.05 in nickel Moth-Eye replicas.

PRACTICAL EXPERIENCE OF USING SELECTIVE SURFACES ON MASS WALL COLLECTORS

D Clarke

David Clarke Associates
4 Tottenham Mews
London W1P 9PJ
UK

1.   SUMMARY

Christopher Taylor Court, a sheltered housing scheme designed by David Clarke Associates for the Bournville Village Trust, incorporates a selective surface mass wall as part of the energy saving design. The paper discusses the practical problems in selecting the appropriate material and incorporating it into a substantial building project with tight economic constraints. The background of Rowheath Solar Village of which Christopher Taylor Court is a part is also described.

2.   BACKGROUND

The Bournville Village Trust was formed in 1900 by George Cadbury for the purpose of administering and developing his model village for the benefit of the residents as a whole. Today, the twelve Trustees, nine of them direct descendants of George Cadbury or his brother Richard, continue to run the 1,000 acre Estate on a self-financing basis, providing the environment for its 22,000 residents according to the Founder's wishes.

The Solar Village at Rowheath is the single largest project of its kind in Western Europe. It comprises nearly 300 dwellings on 7 separate sites, of which Christopher Taylor Court is one, grouped around a 65 acre area of open space. The Village lies on the southern edge of the Bournville Trust Estate and all the dwellings incorporate passive solar techniques in their design and construction, in varying degrees and various ways. (1)

3.   DESIGN BRIEF

The site for Christopher Taylor Court lies between a new development of passive solar family houses and a 17th century timber framed barn. The site is nearly flat and has excellent exposure to the south with views over school playing fields. The brief for the scheme required the provision of 42 flats for elderly people, plus ancillary accomodation. The scheme is funded by the Housing Corporation who require certain standards for this type of sheltered housing:

- all flats must be accessible without using stairs
- the access to all flats must be enclosed and heated to 16 deg C
- the flats must be constantly heated to 21 deg C (compared to average 18 deg C for family housing).

In addition to the minimum functional requirements of the brief Bournville Village Trust and ourselves agreed the following design aims:-

- to maximise the energy saving benefits of solar gain and energy conservation using passive solar design principles
- to achieve a character of building which has a clear identity for the community who will live in it, while respecting the character of Bournville.
- to use the building form to define and shelter outside spaces within the site, so that they can be used by the residents as often as possible.

4.   ANALYSIS OF FORM

Various layouts and building massing were assessed. The layout chosen has a predominently southerly orientation for the flats and has the advantage of daylit corridors; in order to provide views from the kitchens there are internal windows which look out at the corridors and through to the outside. (Fig 1).

This compares with a typical sheltered housing scheme which invariably ends up with an east-west orientation with an artificially lit corridor. Our initial analysis showed that there was no energy penalty in adopting a southerly orientation, and the increase in circulation area resulting from a single loaded corridor, was acceptable. (2)

The spacing of the blocks has been designed so that there is no shadow cast by one block on another at the winter solstice (Fig 2 & 3). The single storey southern wing is at a slightly higher elevation to accommodate the 1m fall on the site; there is a shallow ramp to accommodate this level change inside the building.

5.   PASSIVE SYSTEM ASSESSMENT

After the overall layout has been determined and planning permission obtained, optimisation of the building fabric was undertaken using the accepted European simulation model 'ESP' at the ABACUS unit of the University of Strathclyde in Scotland. Using ESP it was possible to analyse various passive systems in great detail.

A simple direct gain system was simulated for comparative purposes but because of glare problems which are greater with elderly people it was decided to use an alternative system in conjunction with normal areas of conventional glazing in doors and windows. The most promising result initially was for a thermosyphoning air collector, but there were concerns over the complexity and control of this system. As a result of these concerns we simulated a simple single glazed mass wall with a selective surface coating, as studies elsewhere had indicated promising results (3). The resulting annual consumption was modelled at 4013kWh compared to 4062kWh for a thermosyphoning collector, and 12400kWh for a conventional design and 4737kWh for a conventional wall in place of the mass wall collector (figures for 2 flats). (Fig 4)

84.

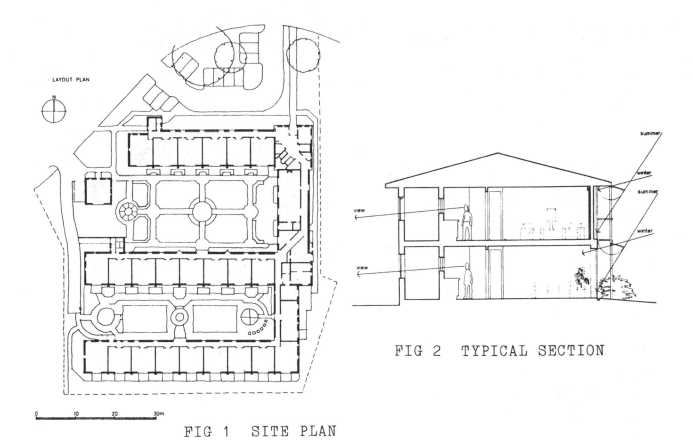

LAYOUT PLAN

N

0   10   20   30m

FIG 1   SITE PLAN

view

view

summer

winter

summer

winter

FIG 2   TYPICAL SECTION

SUMMER

WINTER

SUMMER

WINTER

SUMMER

WINTER

FIG 3   SITE SECTION

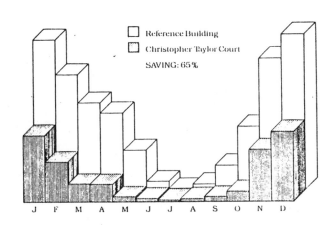

☐ Reference Building
▨ Christopher Taylor Court

SAVING: 65%

J F M A M J J A S O N D

FIG 4 COMPARATIVE HEATING DEMAND

85.

The simplicity of the mass wall persuaded us to adopt this system and look at minor changes to improve the performance.

The final design arrived at using the ESP model for optimisation studies was a combination of direct gain windows with a 200mm dense masonry wall with selective surface and single diffusing glazing. The studies also showed the benefits of movable shades for the direct gain openings (shading of the mass wall did not significantly reduce peak temperatures), and confirmed the effect of internal sliding shutters which are left in front of the mass wall in summer to reduce radiation to the room and in winter are slid across the areas of glazing to increase night time 'U' values.

We also simulated a thinner mass wall (140mm) which would be cheaper to build. There was very little difference (1% increase in fuel requirement) in the winter energy performance but an increase in summertime temperatures. As a result the thickness of the wall was retained at 200mm.

6.   OPTIONS FOR THE CONSTRUCTION OF THE MASS WALL

Having established the optimum physical characteristics of the mass wall for our particular scheme we examined the various construction options.

The innovative part of the wall is of course the selective surface. Fortunately the main producer, Inco Alloys, manufactures these materials in Birmingham and we were able to collaborate with them at an early stage. The choice of products was between Maxorb, a treated nickel foil, and Skysorb, a treated stainless steel plate. The foil is normally used stuck direct to metal surface of flat plate collectors and the stainless steel plate is normally used as the actual collector panel; whereas we were looking for a way of making the selective surface have thermal contact with a wall of dense masonry.

Four options for incorporating these materials were identified:

1.   an insitu concrete wall which could be cast smooth enough to accept Maxorb foil stuck directly on to it.

2.   a prefabricated box constructed of sheet steel with its outer face prepared to take a self adhesive Maxorb foil

3.   pre cast concrete blocks constructed with an outer surface of Skysorb plate of identical dimensions to the blocks used elsewhere in the scheme (440x215x190mm).

4.   a wall of dense concrete blocks constructed on site prepared to take Maxorb foil immediately prior to glazing

Option 1 was quickly ruled out at the cost of the insitu concrete work is very high and would result in the basic massive part of the wall costing many times the cost of Option 4.

86.

Option 2 we investigated in more detail as it seemed a promising prototype for a component which could be used on other sites. The prefabricated box would be relatively light requiring only small cranes to lift and could be manhandled into position.

The box would form a shutter for the concrete filling thus eliminating expensive site shuttering. The quotations obtained for the basic steel box were at a level which ruled out this possibility for this site.

Option 3 was much the most difficult to investigate and led to many meetings with various concrete block manufacturers. The proposal was to construct simple trays of Skysorb stainless steel which would be incorporated in the concrete block manufacturing process, resulting in a selective surface mass wall block which could have wider application.

There are many levels of technology in the concrete block industry. At its most sophisticated and productive the largest companies have enormous machines producing hundreds of concrete blocks per hour with minimal labour, at the other end of the industry there are companies making blocks with simple shuttering on open grounds. It was certainly not possible to consider incorporating steel faces to concrete blocks in the large automated plants; in the hand made type of blocks this certainly was possible but resulted in a cost for the supply of metal faced blocks in excess of £100 per sq.m. (compared to less than £10/sq.m. for a basic unfaced block).

We eventually found a manufacturer who specialised in production of concrete blocks with special facework, usually aggregate of some kind. The plant used was sufficiently adaptable to produce metal face blocks without excessive labour content; metal face plates were made and successful sample blocks produced. The price of a 100mm thick block was quoted at £24/sq.m. which was much more promising but the 190mm thick block which was required for the mass wall was more difficult to make and was quoted at £61/sq.m. And so again costs were too high for this project. There were also some other potential drawbacks in this approach; the face area of the finished wall would be 7% joint material which would not acting as a selective surface, the Skysorb is not such an efficient selective surface (4) and so these two factors would theoretically result in a significant reduction in performance as a selective surface compared to a Maxorb faced wall.

At this point it is useful to put the cost problem in context. In a typical domestic type construction of masonry walls, which is all that the cost limits allow for this type of project, the construction cost of a conventional wall is of the order of £45/sq.m. and it is against this figure that our mass wall should be cost effective, whereas a typical office block facade can easily cost in excessive of £200/sq.m. thus giving much greater scope for flexibility in design; also the wall/floor area ratio is much lower in larger buildings making their total budgets less sensitive to changes in skin design.

The final option of a conventionally constructed wall of solid 200mm concrete blocks is cheaper than the conventional insulated cavity wall parts of the building, the addition of a single glazing and framing mades it roughly equivalent in cost, leaving a target figure of £16/sq.m. (the capitalised value of the mass wall energy savings) for the cost of applying the selective surface.

Earlier applications of selective surface to masonry walls used RTV adhesive troweled onto the masonry wall (5), the foil was then hung vertically in 305mm strips. This method has operated satisfactorily but was laborious to apply and used significant quantities of adhesive. In collaboration with Inco Alloys an alternative was investigated.

The pressure sensitive adhesive used on Maxorb foil for flat plate collectors is a G E Silicones SR 6574 which has been identified as suitable for conditions of weather resistance and temperature.(6) However this is unsuitable for the friable surface of a masonry block or rendered surface, which need a priming layer to stabilise the surface. Small scale tests were carried out by Inco on the silicone weather coat SCM 3304. These tests showed a weak initial adhesion but a build up of bond strength over time to a satisfactory level. (6)

The SCM 3304 system as used by Ruberoid Insulations for many years is a u.v. and weather protection to externally sprayed foam insulation. We were able to use their facilities and equipment to carry out full size trials on a typical section of wall.

Two alternative walls were built, one with smooth faced blocks laid to as even a face as possible, the other a normal block wall with 12mm render coat. Both walls were sprayed with SCM 3304 and after curing (approx 2 hours) the Maxorb foil with a pressure sensitive adhesive was hung.

The trial showed that although the fair faced blocks were suitable for the proposed application, the blocks could not be laid precisely in the same plane resulting in numerous instances where the foil was bridging slight differences of level resulting in a serious loss of thermal contact, whereas on the rendered surface a virtually 100% contact area was achieved. It was therefore concluded that we would have to render the wall and that a possible saving of the render to the wall could not be achieved.

The final type of foil used has the standard adhesive layer and a protective plastic film which could be left on the wall and removed immediately prior to glazing. The foil also has numerous perforations to make it vapour permeable, so that any moisture from the construction can pass through the selective surface layer.

The bottom bead on the glazing to the mass walls has a condensation drain which allows the escape of moisture if it condenses on the inner face of the glass. As it is not a sealed construction this will occur at certain times of year when temperatures fall rapidly at night. The glazing frame also has a

FIG 5 VIEW OF SOUTH WALL

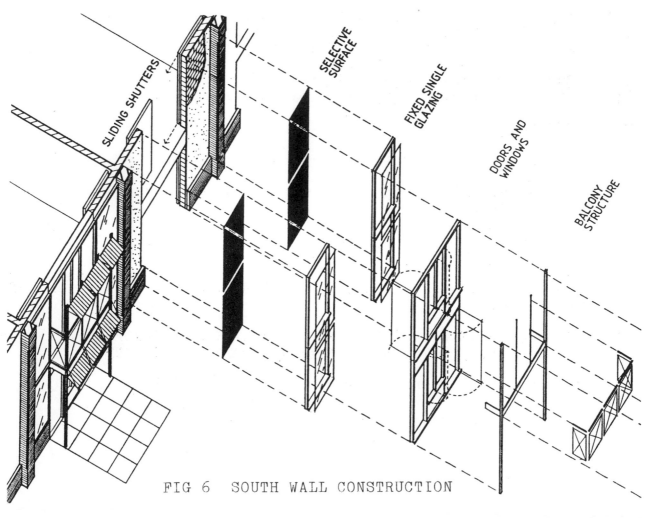

SLIDING SHUTTERS

SELECTIVE SURFACE

FIXED SINGLE GLAZING

DOORS AND WINDOWS

BALCONY STRUCTURE

FIG 6   SOUTH WALL CONSTRUCTION

trickle vent at the head of the frame. This is kept closed in winter and opened in summer months to reduce peak temperatures.

The installation of the selective surface including the initial spray preparation of the wall was undertaken by Ruberoid Insulations. The visual quality of hanging of the foil in site conditions is not as good as in the trial undertaken in a controlled environment, however the diffusing glazing effectively obscures this although at a slight loss in transmission.

7. CONCLUSIONS

Detailed monitoring is being undertaken on 6 flats with probes in the external cavity of the wall, at the outside surface, centre and inside surface, as well as throughout each flat.

Initial measurements show the wall behaving as predicted, but data collection is not yet complete. Analysis of the data will take place later this year.

8. ACKNOWLEDGEMENTS

Funding:
Basic costs: The Housing Corporation
Extra Energy Costs: Bournville Village Trust
and Monitoring   European Commission

Client: Bournville Village Trust
Consultant Architects: David Clarke Associates
Computer Simulations: Abacus, University of
                           Strathclyde
Spec. Subcontractors: Ruberoid Insulation Services
Specialist Supplier: Inco Selective Surfaces Ltd

9. REFERENCES

(1) Proceedings World Solar Congress, Perth 1985.

(2) D Clarke, Proceedings 2nd European Architecture Conference, Munich 1987.

(3) McFarland and Balcomb, 3rd National Passive Solar Conference, San Jose 1979.

(4) Inco Selective Surfaces, published data on properties of Maxorb and Skysorb.

(5) JJ Mason & SJ Adams, Use of Selective Solar Absorber Foils on Trombe Walls, Split Solar Energy meeting, Oct 1987.

(6) JJ Mason & TA Brendel, proceedings SPIE, 1982

ACCELERATED AGEING OF SOLAR ABSORBER SURFACES: TEST METHODS AND DEGRADATION MECHANISMS

P.R. Dolley and M.G. Hutchins
Solar Energy Materials Research Laboratory,
School of Engineering,
Oxford Polytechnic,
Oxford, England.

ABSTRACT

A systematic study of the ageing characteristics, degradation mechanisms and failure modes of selective solar absorbers is being undertaken to assist in the prediction of material service lifetime. This paper presents results for black chrome and nickel pigmented aluminium oxide surfaces following exposure to elevated temperature and controlled temperature-humidity tests respectively. Comparisons were made between results obtained from tests conducted using both high and low loads in order to assess the validity of short duration, high stress level tests for ageing characterisation. The paper concentrates on integrated solar absorptance values as primary indicators of surface response and on measurements of the diffuse and total spectral reflectance for evidence of morphological and compositional change. Surface morphology has been investigated using scanning electron microscopy.

INTRODUCTION

Limited reliable data currently exist on the service lifetime of solar absorber surfaces. Accurate economic benefits are, therefore, difficult to assess. An extensive study of the durability and thermal stability of a wide range of novel and commercially available solar absorber coatings is now in progress. The principal aims of the study are (i) to develop a systematic methodology for the evaluation of absorber coatings using accelerated ageing tests; (ii) to provide input data to mathematical models to predict service lifetime with respect to specific application performance criteria. The work forms part of the U.K. contribution to Task X of the International Energy Agency Solar Heating and Cooling programme. To assist in satisfying the principal aims a number of areas of concern have been identified.

(i)   To identify appropriate accelerated ageing tests.
(ii) To compare degradation mechanisms induced by different tests, e.g.elevated temperature exposure in air and exposure to controlled humidity/temperature environments.
(iii) To compare degradation mechanisms induced by an individual test using different load factors.

(iv)   To examine the suitability of integrated values of
        solar absorptance, α , as primary indicators of
        performance change and both total and diffuse spectral
        reflectance measurements as indicators of degradation
        mechanisms, compositional and morphological change.

(v)   To assess the influence of localised defects and
        imperfections in the long term stability of a surface
        coating (within this paper this point will not be
        discussed).

With respect to (iv) above thermal emittance values could also
be used as primary degradation indicators. We do not consider
such measurements in this paper but these data are being
assembled by us for future evaluation.

Many results have now been obtained for a wide range of
surfaces and some have been previously reported (1,2). In this
paper we present only results for black chrome and nickel
pigmented aluminium oxide absorbers which are relevant to
points (i) - (iv) listed above.

EXPERIMENTAL

Black chrome on nickel on copper samples were obtained from Mti
Solar, U.S.A., and nickel in aluminium oxide samples were
obtained from Sunstrip, Sweden, and Showa, Japan.

All samples were characterised before and after testing. The
standard characterisation procedure used was

(i)   Measurement of near-normal hemispherical spectral
        reflectance, ρ_λ and the diffuse component $\rho_\lambda d$ using a
        Beckman 5240 integrating sphere spectrophotometer
        with a $BaSO_4$ calibration and reference.

(ii)   Calculation of solar absorptance α using 20 selected
        ordinates (3).

(iii)   Examination of surface microstructure by secondary
        electron microscopy using a Jeol JSM 840 instrument.

(iv)   Exposure to defined accelerated ageing test.

(v)   Repetition of (i), (ii) and (iii) above.

Four categories of accelerated ageing test have been employed.

(i)   Exposure to elevated temperatures in air at atmospheric
        pressure as a function of temperature and time.

(ii)   Exposure to constant temperature and constant humidity
        as a function of temperature, humidity and time.

(iii)   Thermal cycling under controlled humidity as a function
        of temperature, humidity and time.

(iv) Outdoor exposure in single-glazed, insulated test
      modules under no-flow stagnation conditions.

For some samples additional measurements were made using
stereo-pair electron microscopy for quantifying surface
roughness parameters and backscattered and secondary electron
microscopy for the estimation of film thickness (1,2).

RESULTS

Table 1 presents the average value of solar absorptance
measured for all test samples of black chrome and nickel in
aluminium oxide in the unexposed condition. All test samples
used had values within 0.01 of the mean.

Thermal cycling and outdoor exposure tests are still in
progress and insufficient data are currently available for a
thorough presentation and interpretation of the results at this
time.

Exposure to elevated temperatures in air at atmospheric pressure

The reference temperature for the elevated temperature tests
was chosen to be 175^0C, a temperature which could reasonably be
achieved under stagnation conditions in collector operation.
All tests carried out for T < 175^0C are referred to as low
stress tests. For the mathematical modelling of service
lifetime an Arhennius model described by Kohl and workers (4)
is being investigated. In its simplest form the model assumes
a single degradation mechanism for the surface with the process
characterised by a single activation energy. For high stress
tests carried out above the reference temperature an
acceleration factor is defined. Measured changes in the
degradation indicator, here changes in solar absorptance $\Delta\alpha$
resulting from high stress tests, are used to scale the data to
an equivalent degradation at an extended time at the reference
temperature. From these derived data service lifetimes can be
estimated against application specific performance criteria,
e.g. that the solar absorptance should not drop below 0.85 say.
This type of model is simple in concept and easy to apply.
However, complexities arise if more than one degradation
mechanism is present or, more importantly, if the degradation
process is temperature dependant and degradation processes not
present in low stress tests are activated under high stress
conditions. It is with this latter point firmly in mind that
the results presented here are considered.

Tables 2 and 3 present the measured changes in solar
absorptance, $\Delta\alpha$, as a function of test temperature and
exposure time for the black chrome and nickel pigmented
aluminium oxide samples respectively. These data may be
interpreted in a number of ways and some examples are now
shown. Figure 1 shows the progressive changes in the spectral
reflectance observed for the black chrome surface following

exposures at 400^0C. By way of comparison Figure 2 presents
spectral reflectance data for black chrome samples exposed to a
range of temperatures when measured values of $\Delta\alpha$ are
comparable.

Figure 3 presents the spectral reflectance of the Sunstrip
Ni-Al$_2$O$_3$ surface for exposures at 300^0, 400^0 and 500^0 C for
those times which result in comparable changes in α. Figure 4
shows a family of spectral reflectance curves for the Showa
Ni-Al$_2$O$_3$ surface in which α progressively decreases.

It is well known that both composition and morphology play
important roles in surfaces which exhibit very high solar
absorptance $\alpha > 0.90$. For the surfaces discussed herein black
chrome is commonly considered to be essentially a cermet of Cr
in Cr$_2$O$_3$ with sub-micron particulate surface roughness. The
role of associated Cr(OH)$_x$ and its effects on optical property
degradation have also been discussed (5). The Ni-Al$_2$O$_3$
surfaces are again cermet materials in which Ni particles are
dispersed within a porous Al$_2$O$_3$ matrix. To assist in
establishing the relative importance of morphological and
compositional changes contributing to observed values of $\Delta\alpha$
we have measured the diffuse component of the spectral
reflectance and examined the surface microstructure by
secondary electron microscopy. Figures 5 and 6 present the
integrated values of the solar reflectance ($\rho = 1 - \alpha$), and its
component diffuse, ρ_d , and specular, ρ_S, contributions for
the black chrome and the Sunstrip Ni- Al$_2$O$_3$ surfaces
respectively.Figures 7-9 are secondary electron micrographs of
black chrome surfaces for unexposed and exposed samples.
Figures 10 and 11 show secondary electron images for unexposed
and exposed Sunstrip Ni- Al$_2$O$_3$ and by way of comparison Figure
12 presents a micrograph of unexposed Showa Ni- Al$_2$O$_3$.
Unexposed samples of black chrome and the Showa Ni- Al$_2$O$_3$
surface were fractured mechanically. Secondary electron images
showing cross-sections of these are shown in Figures 13 and 14
respectively.

Exposure to constant temperature and constant humidity

Results from these tests will not be presented in detail here.
Some results have previously been reported (2,6). Essentially,
for the test conditions employed, black chrome surfaces remain
unchanged. For Ni-Al$_2$O$_3$ surfaces some important changes have
been evidenced. Film thickness increases have been observed
following exposure to moisture and this is attributed to the
take up of water into the film, possibly as aluminium
hydroxide, Al(OH)$_3$. Representative changes in the spectral
reflectance of exposed Ni-Al$_2$O$_3$ surfaces are shown in Figure
15, and the surface microstructure of an exposed surface is
shown in Figure 16.

DISCUSSION

For the methodology of accelerated ageing currently being
developed degradation mechanisms ideally should be temperature
independent allowing short duration, high stress level tests to
be performed to augment tests carried out under reference

conditions for the longest necessary exposures. Thus the
primary consideration is whether the high stress tests induce
similar surface changes when results are compared with long-
term low stress level exposures. Figure 2 shows that spectral
reflectance curves obtained for black chrome of $150^\circ C$, $300^\circ C$,
$400^\circ C$ and $500^\circ C$ exhibit similar features and are all different
to the control sample, i.e. change has occurred. The prominent
change is the appearance of humps in the spectral reflectance
curve most notably at ~ 0.8 μm and also present ~ 0.5 μm. These
result from the crystal field spectra of Cr_2O_3 which is formed
by the oxidation of Cr particles under the elevated temperature
conditions. Similar spectral reflectance changes have also
been observed in nickel and cobalt blacks (7,8). In addition
to compositional changes the surface microstructure of black
chrome is altered by elevated temperature testing and Figure 5
shows that the diffuse contribution to the total reflectance of
black chrome exposed at $400^\circ C$ in air increases with exposure
time. Similar behaviour is observed at other temperatures.
The specular component of the reflectance remains largely
unchanged. Morphological changes are also evident in the
secondary electron images of Figures 7-9. Techniques for
quantifying surface microroughness parameters by stereo-pair
electron microscopy have been developed (1,2) which enable the
effects of surface roughness on optical properties to be
quantitatively assessed (9). Results of this work will not be
discussed here. Figure 9 reveals the appearance of relatively
large crystals, believed to be Cr_2O_3, which are only evident in
surfaces which have been exposed at $500^\circ C$ in air for periods of
time greater than 3 hours. This we attribute to the presence
of a degradation process which is not activated at the lower
temperatures investigated and thus restricts the range of
loads which can be applied to the black chrome surface for the
purpose of accelerated ageing testing.

In contrast the responses of $Ni-Al_2O_3$ surfaces at $500^\circ C$ are
broadly consistent with data acquired at lower temperatures.
Figure 3 shows spectral reflectance curves for Sunstrip $Ni-
Al_2O_3$ surfaces which show similar changes in α under different
exposure conditions of temperature and time. The film
thickness is essentially unchanged. This behaviour is also
confirmed in the Showa $Ni-Al_2O_3$ surface and is clearly evident
in Figure 4 which shows the progressive change in ρ_λ for
elevated temperature tests. Film thickness is essentially
constant, but the films become increasingly transparent and the
amplitude of the interference fringes increases as the load
increases. The proposed degradation mechanism here is the
oxidation of Ni to, say, NiO resulting in a decrease in k, the
extinction coefficient, for the film. The rate of oxidation of
Ni at low stress levels may not be a serious problem. The
surface microstructure of this class of surfaces does not
differ markedly following exposure to elevated temperatures
(Figs 10-12) and the ratio of the diffuse to specular
components of the solar reflectance (Figure 6) remains
essentially constant. $Ni-Al_2O_3$ surfaces are observed to
respond in a different manner when exposed to humidity and
moisture. When exposed to moisture α decreases and film
thickness increases. In addition, the microstructure is

altered. Hence the surface response is test dependent. A better understanding of test dependent surface response is necessary if (i) combined test methods are to be used in accelerated ageing studies, (ii) surfaces which have been exposed to in-service conditions are to be properly evaluated.

CONCLUSIONS

At this stage it seems reasonable to continue performing high stress, short duration accelerated ageing tests to extend data acquired from low stress tests for the purposes of predicting service lifetime and identifying degradation mechanisms. Care must be taken with this approach. Different surfaces respond in different ways to individual tests and furthermore the response of a coating such as black chrome to an individual test may also be dependant upon the substrate employed. Large numbers of individual tests need to be performed and this is expensive in time for both the experimental and the analytical aspects of the work. Future reports will also contain results for other surface types.

REFERENCES

1. HUTCHINS, M.G., DOLLEY, P.R. and LLOYD, G.W., Durability assessment and microstructural characterisation of selective solar absorber surfaces, Proc. IEA SHC Task X Workshop on Material Demand, pp 97-106, Tokyo, Japan, 1987.
2. DOLLEY, P.R. and HUTCHINS, M.G., Accelerated ageing and durability testing of spectraly selective solar absorber surfaces, Proc. ISES Solar World Congress 1987, Hamburg, Pergamon (in press).
3. ASTM E891-82, Standard for Terrestrial Direct Normal Solar Spectral Irradiance Tables for Air Mass 1.5, pp 692-699, 1982.
4. KOHL M., GINDELE, K. and MAST M., Accelerated ageing tests of copper oxide and Ni - MgF_2 cermet solar absorber coatings, Solar Energy Materials 16, 155, 187.
5. ZAJAC, G., SMITH, G.B. and IGNATIEV, A., Refinement of solar absorbing black chrome microstructure and its relationship to optical degradation mechanisms, J. Appl. Phys. 51 (10), 5544, 1980.
6. DOLLEY, P.R. and HUTCHINS, M.G., Accelerated testing of selective solar absorber surfaces : 2 Exploratory study of surface stability to a controlled temperature-humidity environment, Solar Energy Materials Research Laboratory Report No. 87/7, Oxford Polytechnic, 1987.
7. HUTCHINS, M.G., WRIGHT, P.J. and GREBENIK, P.D., Comparison of different forms of black cobalt selective solar absorber surfaces, Solar Energy Materials 16, 113, 1987.
8. COOK, J.G. and KOFFYBERG, F.P., Solar thermal absorbers employing oxides of Ni and Co, Solar Energy Materials 10, 55, 1984.
9. KOHL, M. and GINDELE, K., Determination of the characterising parameters of rough surfaces for solar energy conversion, Solar Energy Materials 16, 167, 1987.

ACKNOWLEDGEMENTS

This work is supported by a grant from the Science and
Engineering Research Council. The authors express their thanks
to all participants in the Task X absorbers study of the IEA
Solar Heating and Cooling Programme and in particular to Drs M
Kohl and K Gindele of the University of Stuttgart for their
helpful advice and support.

Sample	α as prepared
Black Chrome	0.96
Sunstrip Ni-Al$_2$O$_3$	0.90
Showa Ni-Al$_2$O$_3$	0.92

Table 1: Solar absorptance values for unexposed samples of Mti black chrome, Sunstrip Ni-Al$_2$O$_3$ and Showa Ni-Al$_2$O$_3$ absorber surfaces.

Temperature (^{O}C)	Time (h)	$\Delta\alpha$
150	100	-0.01
150	856	-0.04
200	1	-0.02
200	200	-0.02
300	1	-0.01
300	10	-0.02
300	200	-0.02
400	3	-0.03
400	6	-0.04
400	10	-0.06
400	20	-0.07
400	30	-0.09
500	1	-0.08
500	3	-0.16
500	6	-0.17
500	10	-0.17
500	30	-0.16

Table 2: Measured changes in the solar absorptance, $\Delta\alpha$, as as a function of test temperature and exposure time for Mti black chrome.

Temperature ($^{\circ}C$)	Time (h)	$\Delta\alpha$
SURFACE: Sunstrip		
300	12	−0.03
300	120	−0.05
400	3	−0.05
400	10	−0.06
400	12	−0.07
400	50	−0.05
500	1	−0.04
500	3	−0.12
500	6	−0.09
500	10	−0.08
500	30	−0.12
SURFACE: Showa		
400	3	−0.02
400	10	−0.03
400	20	−0.03
400	50	−0.08
500	1	−0.07
500	3	−0.07
500	6	−0.08
500	10	−0.06
500	300	−0.10

Table 3: Measured changes in the solar absorptance, $\Delta\alpha$, as a function of test temperature and exposure time for Sunstrip and Showa $Ni-Al_2O_3$ samples.

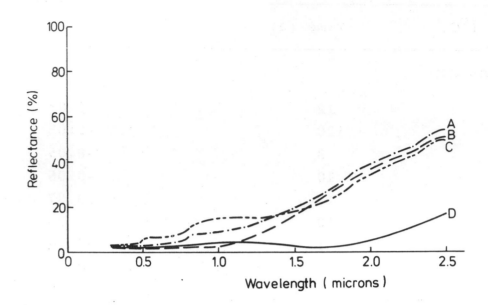

Figure 1: Spectral reflectance of black chrome surfaces following elevated temperature exposure at 400°C A 20h, B 3h, C 50h, D unexposed.

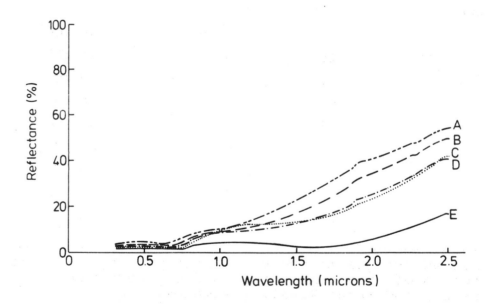

Figure 2: Spectral reflectance of black chrome surfaces following elevated temperature exposure, $-0.08 \leq \Delta\alpha < -0.04$, A 500°C, 1 h; B 400°C, 10h; C 150°C, 856h; D 300°C, 200h; E unexposed.

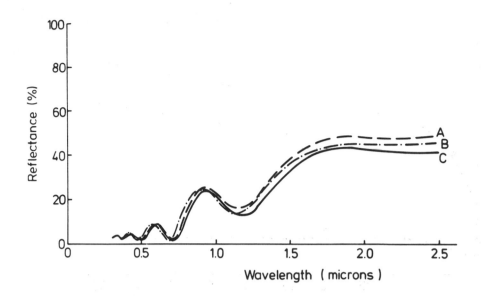

Figure 3: Spectral reflectance of Sunstrip Ni-Al$_2$O$_3$ surfaces following elevated temperature exposure, $-0.05 < \triangle\alpha < -0.04$, A 500°C, 1h; B 400°C, 3h; C 300°C, 120h.

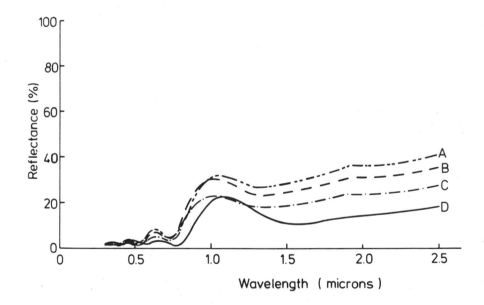

Figure 4: Spectral reflectance of Showa Ni-Al$_2$O$_3$ surfaces following elevated temperature exposure, $-0.08 \leq \triangle\alpha < -0.03$, A 500°C, 6h; B 500°C 1h; C 400°C, 1h, D unexposed.

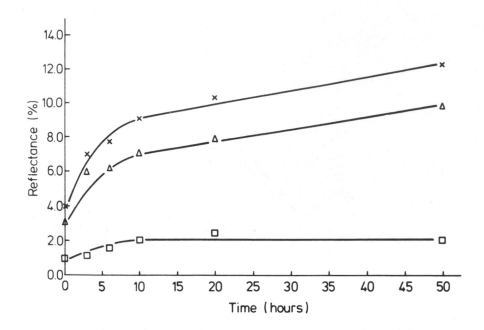

Figure 5:   Integrated values of solar reflectance (x) and the associated diffuse (△) and specular (▢) components for black chrome surfaces following elevated temperature exposure at 400°C for various times.

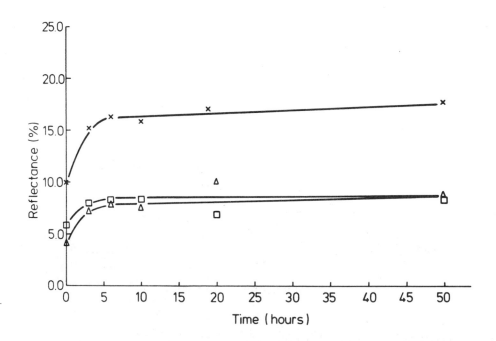

Figure 6:   Integrated values of the solar reflectance (x) and the associated diffuse (△) and specular (▢) components for Sunstrip Ni-Al$_2$O$_3$ surfaces following elevated temperature exposure at 400°C for various times.

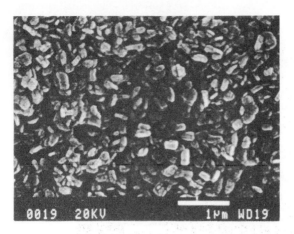

Figure 7: Secondary electron micrograph of black chrome as prepared, α = 0.96.

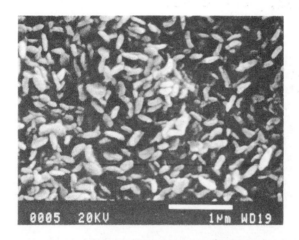

Figure 8: Secondary electron micrograph of black chrome following elevated temperature exposure at 400°C for 50h, α = 0.88.

Figure 9: Secondary electron micrograph of black chrome following elevated temperature exposure at 500°C for 29h, α = 0.81.

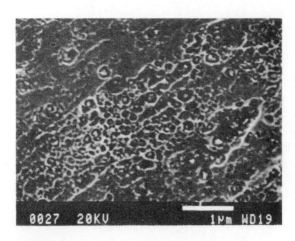

Figure 10: Secondary electron micrograph of Sunstrip $Ni-Al_2O_3$ as prepared, $\alpha = 0.90$.

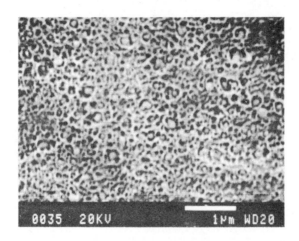

Figure 11: Secondary electron micrograph of Sunstrip $Ni-Al_2O_3$ following elevated temperature exposure at $500^{\circ}C$ for 29h, $\alpha = 0.78$.

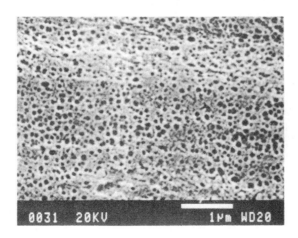

Figure 12: Secondary electron micrograph of Showa $Ni-Al_2O_3$ as prepared, $\alpha = 0.92$.

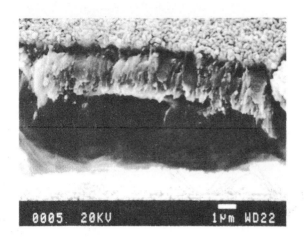

Figure 13: Secondary electron micrograph of a fractured cross-section of unexposed black chrome.

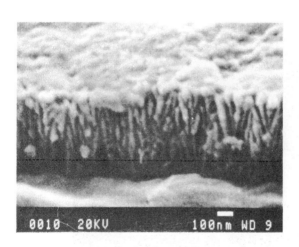

Figure 14: Secondary electron micrograph of a fractured cross-section of unexposed Showa Ni-Al_2O_3 showing the porous nature of the Al_2O_3 host matrix.

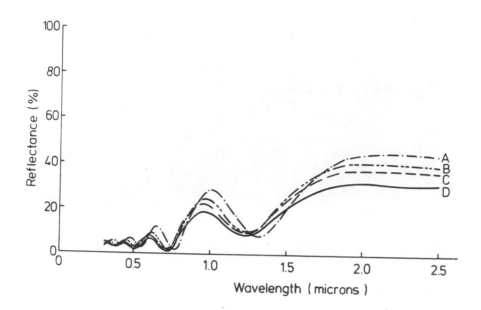

Figure 15: Spectral reflectance of Sunstrip Ni-Al$_2$O$_3$ surfaces following exposure to 90°C and 95% RH for A 168h, B 6h, C 3h, D unexposed.

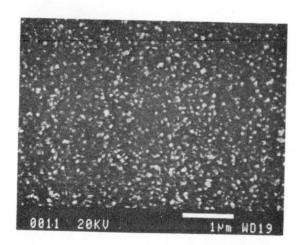

Figure 16: Secondary electron micrograph of Sunstrip Ni-Al$_2$O$_3$ following exposure to 90°C and 95% RH for 720h. Small particles are seen to cover much of the surface.

DURABILITY TESTING AND SERVICE LIFETIME PREDICTION OF SOLAR ENERGY MATERIALS

M. Köhl and K. Gindele

Institut für Theorie der
Elektrotechnik,
der Universität Stuttgart
Pfaffenwaldring 47,
D 7000 Stuttgart 80
F.R.Germany

U. Frei

Interkantonales Technikum
Rapperswil,

Oberseestraße 10
CH 8640 Rapperswil
Switzerland

INTRODUCTION

In the past, great efforts have been made to develop and to understand selective solar absorber coatings. The number of coatings with reasonably good optical properties and the collected knowledge of their physics show obviously the success of these efforts. But, what we are seriously missing today is a better understanding of the longtime behaviour of these coatings and of their ageing processes. It is not an easy task to investigate ageing mechanisms comprehensively and to develop procedures which enable to estimate the lifetime of an absorber under given service conditions.

There is some published work on ageing of absorbers under temperature loads, but only few research work has been done, to study the influence of high humidity on selective absorber coatings. In many flat plate solar collectors however, condensation occurs on the inner side of the cover which is a clear evidence that the atmosphere inside of the collector must be very humid. Unfortunately the exact environmental conditions are not very well known. Nevertheless, this frequently observed condensation phenomenon shows the importance of studying in detail what happens with selective absorber coatings exposed to humid conditions.

ACCELERATED AGEING TESTS

Every product in use degrades in the course of time by getting worn out or by weathering. Its durability should be good enough to ensure an acceptable lifetime, at which the performance of the product falls beyond the performance requirements. Reliable informations about the durability of a product can only be yielded after the real use during the expected lifetime. This way is not acceptable, however, for the development of new products with a long lifetime, for instance solar absorbers. In this case accelerated short-term tests are needed. Suitable accelerated durability tests must take into account the specific in-use conditions and the specific material composition of the test samples. This paper will focus on non-polymeric solar absorber coatings used in glazed flat plate solar collectors with their specific working conditions, and consequently arising stresses of the absorber.

Degradation Factors

Indoor durability tests of the absorber should be carried out independently of the individual outdoor stress conditions, but should take into account all possible natural stresses which depend qualitatively and quantitatively on the collector type, the system design, and the local climate. One method is the separation of the natural stress into single **degradation factors.** The influence of the single degradation factors can be investigated more or less seperately and reproducibly by laboratory tests in an appropriate parameter range independent on individual applications, in order to identify specific sensitivities of the absorber.

Unavoidable degradation factors at photothermal energy conversion are the absorber temperature, the cyclic temperature variation and the solar irradiation. The temperature load depends on the local climate, the performance, the operation conditions of the whole system, and the heat transfer from the absorber to the fluid. The maximum temperature is reached under stagnation conditions. The frequency and the amplitude of temperature cycles depend on the same parameters and vary seasonally.

The solar irradiation primarily causes the temperature load, but moreover photolytic degradation could occur. The transmission of the glazing as well as the concentration factor and the spectral transmittance of a concentrating system influence the grade of this load.

In non-evacuated collectors, components of the atmosphere surrounding the absorber can act as reaction partners in degradation processes. Mainly oxygen, humidity and moisture, as well as pollutants like salt, sulfur-dioxide, and nitrogen-oxides must be considered as degradation factors. Their concentrations depend on the collector design and the local climate.

Combinations of different degradation factors, which may cause additional synergetic effects, are difficult to realize as accelerated tests in laboratories in a suitable way, but can easily but unreproducibly obtained at outdoor exposure to natural weathering at in-use conditions or slightly accelerated at stagnation conditions.

Acceleration

The goal of accelerated durability tests is to obtain the same degradation of the sample as after the lifetime under in-use conditions, but in a very much shorter time. This is usually done by a quantitative enhancement of the degradation factor applied in the test. The correlation between this enhancement and the acceleration of the degradation can be used for an extrapolation to the in-use level of the degradation factor in order to estimate the degradation at in-use conditions.

For this, and for the design of accelerated ageing tests, further knowledge of the degradation mechanisms is very helpful.

Degradation Modes

The effect of a degradation factor on an absorber depends on its chemical and physical composition. Selective solar absorber coatings can be realized according to several physical principles /1,2/ and with numerous materials (see e.g. /3, 4/) deposited on various substrates. The degradation factors acting on the absorber yield specific ageing processes, the so-called **degradation modes**, which changes the chemical and/or physical composition of the absorber.

A coarse classification of the degradation modes can be made as follows: In evacuated (or inert gas filled) collectors the temperature load might accelerate internal diffusion processes, internal chemical reactions between different components of the absorber, or desorption of volatile matter (outgassing), and might cause recristallisation or phase transitions. Additional mechanical degradation (microcracks, blistering) might occur due to the cycling of the temperature. The solar irradiation firstly causes a temperature load with the respective degradation modes, but can moreover accelerate these processes or set up new ones by photocatalysis.

An environmental atmosphere can provide external diffusion and external chemical reactions, in addition.

A combination of the different degradation factors (at the natural ageing) may yield synergetic effects leading to other degradation modes than the single degradation factors applied seperately.

Acceleration Factor Equations

The degradation modes are a key point for the nature of accelerated ageing tests. The fundamental physical laws ruling the degradation modes describe the relation between the degradation factor and the time. Therefore, they can be used for the evaluation of the acceleration effect of enhanced loads, which can be described by an acceleration factor, defined as the ratio between the time interval t_r at a reference load level and the time t_n needed at another load level for obtaining the same degradation of the observed property:

$$a_n = \frac{t_r}{t_n} \tag{1}$$

The degradation caused by diffusion processes, chemical reactions or desorption is accelerated by an enhanced temperature in a similar way: The temperature dependence of the diffusion constant as well as that of the reaction velocity obeys a law of an Arrhenius-type /6,7/, which leads to the same equation for the acceleration factor a_n for a reference temperature T_r, and the load levels T_n for obtaining the same degradation.

$$a_n = \exp\left(\frac{\emptyset}{R}\left(\frac{1}{T_r} - \frac{1}{T_n}\right)\right) \tag{2}$$

where $\emptyset$ is the activation energy of the respective process and R is the gas constant. This relation is often used in durability testing /8/ and was already applied to solar absorber ageing tests, too /7,9/.

An acceleration of the degradation caused by parts of the atmosphere (namely the degradation modes "external diffusion" and "external chemical reactions") can be achieved either by an increase of the temperature or by an increase of their concentration. The reaction velocity of chemical reactions of the first order or of equilibrium reactions, as well as the diffusion current density at diffusion processes, is directly proportional to the initial concentration C, yielding the acceleration factor equation:

$$a_n = \frac{C}{C_0} \tag{3}$$

An increase of the concentration is a suitable method for the investigation of the influence of pollutants occuring in very small concentrations, but acceleration by an increase of the temperature is nessessary for tests with humidity and moisture, which can occur with highest concentrations in collectors, or oxygen, for which a high increase of the concentration cannot be achieved. A similarity to the effect of concentrations of external reactants is shown by the irradiance I causing photolytic reactions. Their acceleration factor

$$a_n = \left(\frac{I}{I_0}\right)^A \tag{4}$$

depends on a material constant A, which has values close to 1 /6/.

Comprehensive overviews on acceleration factor equations, also called time transformation functions, can be found in /6/ and /8/. The nature of phase transitions and recrystallisation processes leads to serious problems not only in designing accelerated tests for these degradation modes, but mainly because they can disturb the applicability of accelerated ageing tests for the other modes.

The temperature plays an outstanding role in accelerated ageing tests. First, very high acceleration factors can be achieved by small variations of the temperature due to the exponential law in equation (2). A temperature rise from 200°C to 300°C yields an acceleration by a factor of about 100 to 10000 for activation energies from 100 kJ/mol to 200 kJ/mol (see figure 1). The simulation of the ageing effect of temperature loads of 80°C for 10 years or 200°C for 1 year could be carried out in one week short-term tests at temperatures between 100°C and 200°C, or between 220°C and 300°C, respectively (see figure 2), depending on the activation energies of the degradation modes. For the evaluation of the activation energies, however, different tests at various temperatures are required. Second, the temperature is also an important degradation factor for most of the degradation modes. Therefore a controlled temperature load in addition to other degradation factors helps accelerating the tests, but on the other hand attention has to be paid to the sample temperatures by performing ageing tests with other degradation factors, especially with irradiation.

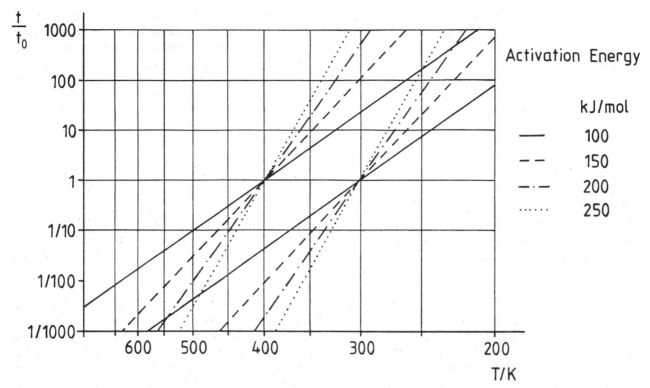

Fig. 1:   Equivalent testing times (normalized to 400°C) as
            functions of the test temperature for different
            activation energies.

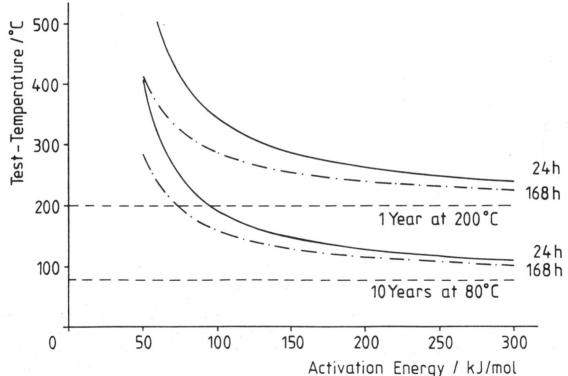

Fig. 2:   Equivalent test temperatures as functions of the
            activation energy for a one day or a one week
            accelerated temperature test simulating 10 years
            exposure to 80°C and 1 year exposure to 200°C,
            respectively.

Usually the degradation modes are unknown and also not detected during the ageing tests. The changes of the materials composition depending on degradation factors, degradation modes and time are normally not observed, too. The physical properties, however, which are finally the points of interest, can be measured as a function of time during the exposition to a constant load, if this load causes changes of the materials composition and if these changes influence the observed property in a measurable way. The measured values - the solar absorptance for instance - form the **degradation function** of this property for the investigated absorber and the applied degradation factor level. Tests at other levels yield other exposition times for the same amount of degradation. Relating these exposure times at different stress levels to the degradation function (if this function is monotonous in time) allows the evaluation of the acceleration factor according to equation (1) as a function of the stress level.

Now the appropriate time transformation function has to be found: Equation 2, 3, 4 or, may be, another one (see e. g. /6/ /8/), If no good correlation for these so-called mechanistic models could be found, it can be assumed that either degradation modes are going on, which cannot be described by these acceleration factor equations (e. g. phase transitions), or different degradation modes are acting simultaneously. In such cases non-mechanistic approaches /6/ can be used, which do not require a knowledge of the degradation mechanisms. The acceleration factors can be approximated as a function of the degradation factor by means of a mathematical best fit function. If the evaluation of the acceleration factors is not possible, the degradation curve itself can be approximated by a model with parameters depending on the degradation factors (see e. g. /6/).

The degradation function can be transformed to any degradation factor level, especially to that at working conditions, by means of the evaluated time transformation function, which describes the dynamic of the degradation process. The lifetime, which is the ageing time (in this case only for one degradation factor) until the degradation limit (appointed by the performance requirements) is reached, can now be read directly as the abszissa of the degradation limit in the transformed degradation function.

This method presumes, of course, that the degradation modes will be the same for the low stress levels at working conditions, as for enhanced test conditions. This assumption can only be justified by a comparison with samples aged naturally under monitored working conditions. For some degradation factors (e. g. temperature, irradiation) it should be plausible to postulate, that at lower stress levels no severer degradation processes run than at higher stress levels.

TEST PROCEDURES

The degradation as a function of time was determined by interrupting the tests from time to time for measurements of the optical properties. This procedure allows the observation of the proceeding degradation at the same sample and saves time or equipment, but has the disadvantage of the interruptions of the ageing process, in contrast to another possible procedure, which requires always another sample for each time interval.

The temperature tests were carried out in two circulating air ovens, with temperature ranges from room temperature up to 750°C. The samples were exposed, electrically isolated, to constant temperatures in a range between 250°C and 600°C, depending on the respective degradation of the solar absorptance. For the optical measurements the ovens and the samples cooled down slowly, in order to avoid thermal shocks.

The humidity tests were carried out at 90°C and 95 % r. h. in a commercially available humidity cabinet.

For the condensation tests the samples were mounted electrically isolated, but with good thermal contact, on a special fluid-cooled sample holder. They were cooled to 5 K below the temperature of the humidity cabinet at 95 % r. h., which yielded a permanent condensation on the sample surfaces.

The outdoor tests, performed under stagnation conditions with commercial flat plate collectors in Rapperswil, Switzerland, are described in the paper "Innovative test procedures for assessment of the durability of solar collectors" at this conference.

The near-normal/hemispherical reflectance reported in this paper have been measured spectrally by means of Zeiss PMQ spectrophotometers equipped with $BaSO_4$-coated integrating spheres (wavelength range 0,36µm-2,5µm) and by means of Bruker Fourier transform spectrometers equipped with diffuse-gold coated integrating spheres (2µm-15µm).The solar absorptance and the thermal emittance were calculated by convoluting the spectral data with the solar spectrum AM 1.5, and with the Planck funktion for a radiator temperature of 373K, respectively. A detailed discussion of the measurement technique in the infrared is given in the paper "Spectral measurements of the infrared reflectance" at the same conference.

RESULTS AND DISCUSSION

The methods described above are now to be illustrated by test results obtained for a black nickel coating, Maxorb (produced by Inco, UK), and two nickel-pigmented aluminium- oxide absorbers, Sunstrip (manufactured by Gränges, Sweden) and Evidal (manufactured by VDM, FRG). A detailed discussion of these tests will be published soon /10/. Some test results are compiled in table 1.

The degradation curves for the solar absorptance α_s (AM 1.5) obtained by temperature tests (see figure 3 - 5) were used for a lifetime time estimation according to equation 2. The Arrhenius-

	unaged		outdoor		temperature 50 h at 400 °C		95 % humidity 650 h at 90 °C		condensation 650 h at 45 °C	
	α_S	ε	α_S	ε	α_S	ε	α_S	ε	α_S	ε
Maxorb	.939	.09	.947	.09	.950	.06	.942	.07	.955	.09
Sunstrip	.924	.16	.917	.17	.894	.14	.915	.26	.921	.84
Evidal	.915	.18	.893	.15	.907	.12	.915	.22	.926	.84

Table 1: Optical properties α_S (AM 1,5) and directional emittance ε (100°C) of the tested sample.

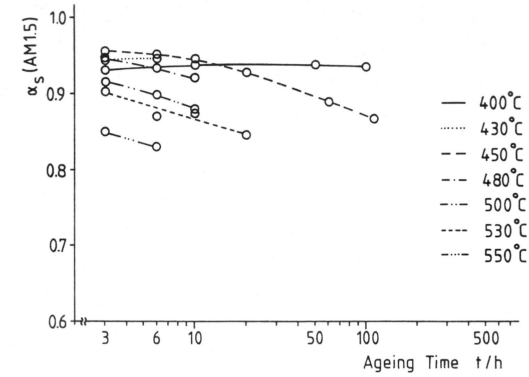

Fig. 3: Solar absorptance of the Maxorb coating as a function of the test time at different constant temperature tests.

plots of the acceleration factors show a good correlation with a straightline, demonstrating the suitability of this acceleration factor equation (see figure 6). The evaluated activation energies are very similar for all coatings: Maxorb 202 kJ/mol, Sunstrip 188 kJ/mol and Evidal 193 kJ/mol. A decrease of 5 % of the absorptance after 10 years exposure to a constant temperature was estimated for temperatures of 320°C for Maxorb, 310°C for Sunstrip and 330°C for Evidal. The thermal emittance decreased during the temperature tests slightly for all coatings. Therefore no problems with the temperature stability of these coatings used in flat plate collectors could be expected, apart from that of the adhesive of the Maxorb foil, which completely peeled off during all temperature tests.

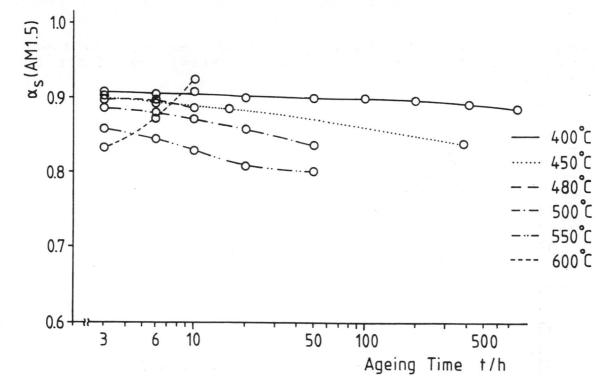

Fig. 4:   Solar absorptance of the Sunstrip coating as a function
          of the test time at different constant temperature
          tests.

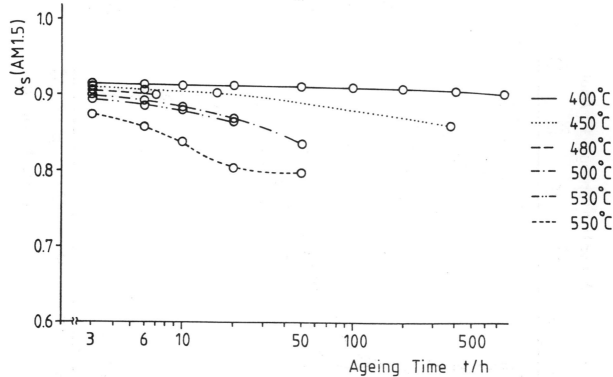

Fig. 5:   Solar absorptance of the Evidal coating as a function
          of the test time at different constant temperature
          tests.

The acceleration factor method applied to the **spectral**
reflectance in the solar range yielded corresponding values for
the activationenergy for the Sunstrip and the Evidal coatings in
that spectral range, where remarkable changes of the reflectance
occurred (see figures 8 and 9). On the contrary the Maxorb
coating showed three regions with different activation energy

ranges, indicating different ageing processes, but the range with the highest activation energy is dominating the effect on α_s (see figure 7).

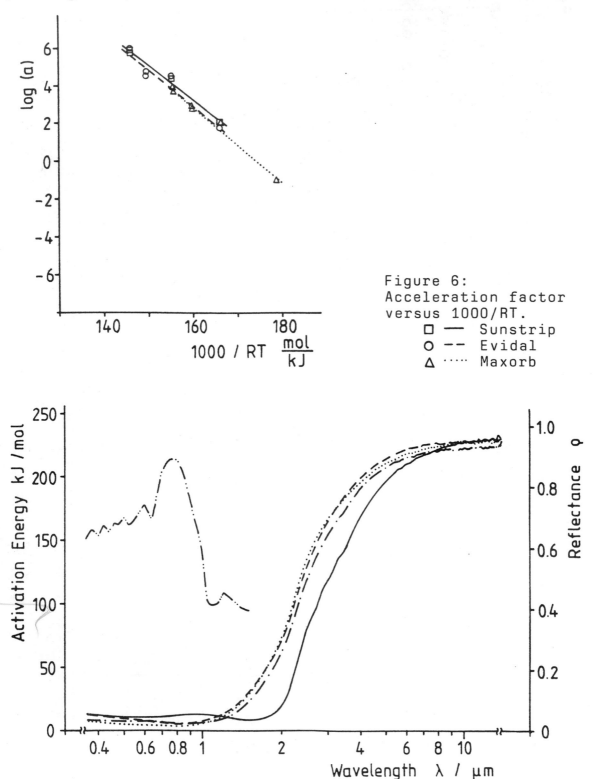

Figure 6:
Acceleration factor versus 1000/RT.
□ —— Sunstrip
○ – – Evidal
△ ····· Maxorb

Fig. 7:   Spectral near-normal/hemispherical reflectance of the tested Maxorb coatings: —··— activation energy
—— unaged,   – – temperature test 10 h at 400°C
····· outdoors,   —·— condensation test 650 h at 45°C

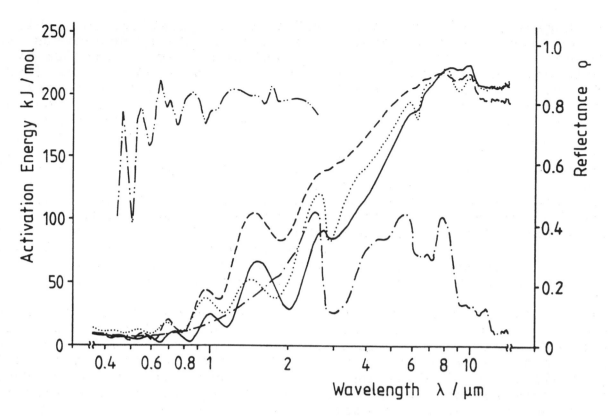

Fig. 8: Spectral near-normal/hemispherical reflectance of the tested Sunstrip coatings: —··— activation energy
—— unaged, —— temperature test 20 h at 500°C
······ outdoors, —·— condensation test 650 h at 45°C

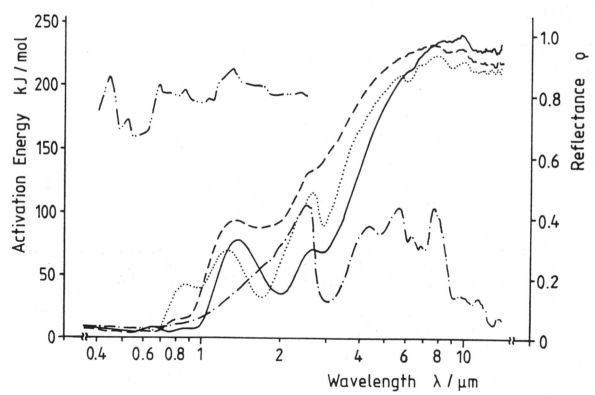

Fig. 9: Spectral near-normal/hemispherical reflectance of the tested Evidal coatings: —··— activation energy
—— unaged, —— temperature test 20 h at 530°C
······ outdoors, —·— condensation test 650 h at 45°C

The humidity test showed low effect on all coatings (s. table 1) in contrast to the condensation tests, which detoriated the thermal emittance of the nickel-pigmented aluminium-oxide coatings. The spectra of the Sunstrip (figure 8) and the Evidal (figure 9) coatings show after the condensation test nearly the same water absorption bands (at 3 µm and 6 µm) and aluminium-hydroxide bands (above 9 µm). The same absorption bands, but less pronounced, can be found in the spectra of the outdoor-tested Evidal and Sunstrip (s. figure 8 and 9),indicating degradation due to moisture. The naturaly aged Sunstrip sample was part of a collector in a DHW-system working under unknown conditions in the Netherlands (provided by B. Brouwer, Holland Solar). The Evidal coating was tested in an inertgas filled collector under stagnation conditions for 2 years in Rapperswil, but the glazing became defect and condensation in the collector occurred. For the Maxorb coating the same small changes of the spectral reflectance could be found after all tests (figure 7).

These examples show the usability of accelerated indoor tests for the investigation of specific sensitivities of absorber coatings. The estimation of the long term temperature stability can be carried out by means at a mechanistic acceleration factor model, generally. The application of such models for moisture tests needs more research work in future, as well as the investigation of other degradation factors, like irradiation, pollutants and cyclic tests. Another important task is the development of suitable methods for the comparison of the effects of natural and accelerated ageing.

REFERENCES

/1/ A. B. Meinel, M. P. Meinel,
     "Applied Solar Energy", Addison Wesley (1976).
/2/ B. O. Seraphin,
     "Thin Films in photothermal energy conversion",
     Thin Solid Films, 90 (1982) 295-403.
/3/ W. F. Bogaert, C. M. Lampert,
     "Materials for photothermal solar energy conversion",
     J. Mat. Science, 18 (1983) 2847-2875.
/4/ G. A. Niklasson, C. G. Granqvist,
     "Surfaces for selective absorption of solar energy: an
     annotated bibliography 1955-1981",
     J. Mat. Science, 18 (1983) 3475-3534.
/5/ A. W. Czanderna,
     "Surface and interface studies and the stability of solid
     solar energy materials", in Solar Materials Science,
      ed. L. E. Murr, Academic Press, New York, 1980, 93-147.
/6/ B. Carlsson ed.,
     "Service Life Prediction Methods for Materials in Solar
     Heating and Cooling",
     Report of IEA/SHCP Task X, to be published 1988.
/7/ M. Köhl, K. Gindele, M. Mast,
     "Accelerated ageing tests of copper-oxide and Ni-MgF$_2$-cermet
     solar absorber coatings",
     Solar Energy Materials 16 (1987) 155-166.

/8/ J. W. Martin,
"Time transformation fuctions commonly used in life testing
analysis".
Durability of Building Materials, 1 (1982) 175-194.
/9/ R. B. Pettit,
Accelerated temperature ageing of Black Chrome solar
selective coatings",
Solar Energy Materials 8 (1983) 349-361.
/10/ M. Köhl et al., to be submitted to Solar Energy Materials.

Innovative test procedure for assessment of the durability of solar collectors

===

U.Frei & T.Häuselmann

Solar Energy Laboratory, LBS
School of Engineering, ITR
CH-8640 Rapperswil, Switzerland

1. Introduction

Qualification tests procedures have been developed for many
years by several research laboratories and international organi-
zations like the Commission of the European Communities "CEC" or
the International Energy Agency "IEA". Most of the proposed approaches consist of a number of single tests performed indoor
or outdoor. Usually the test period is only a few hours up to
several days e.g. for an outdoor stagnation test. The tests are
usually performed only once; that means ageing effects of a
certain test, which might be only of importance during an other
test, can of course not be detected. One example: it is of great
importance whether to perform first the high temperature test,
followed by the rain penetration test or vice versa. Also many
ageing effects like small leaks, ageing of sealants, gaskets,
plastic covers etc. can only be detected after several months of
operation.
The primary philosophy of the here presented procedure was to
include every possible load condition, which can occur under in
use conditions, repeated continuously during a specified testing
time. To achieve a certain acceleration compared with the real
conditions the test last 24 hours a day, that means the collectors are heated and cooled by respective heating or cooling
devices. Instead of applying an enormous solar simulator and a
climatic cabinet, the test is performed outdoors.
This new test procedure will tell, if a collector is able, under
certain boundary conditions, to survive a lifetime of 15 years
without major defects or a significant decrease in performance.

This test is developed only for flat plate collectors used in closed loops, with glycol water mixtures as heat transfer fluid and a maximum stagnation temperature of 220°C.

2. Test description

Each test place consists of two collectors of the same product. They have to be mounted on the outdoor collector test facility according to the specifications of the manufacturer. If there are any fittings or joints available from the collector manufacturer, they have to be used according to the manufacturer's installation guide.
To evaluate an appropriate test program, we have to analyse all possible operational conditions combined with the possible weather conditions:

1. Normal operating condition in the temperature range from 20°C to 80°C
2. Stagnation condition (no flow through the collector)
3. External shock (thunderstorm after stagnation condition)
4. Internal shock (system start after stagnation condition)
5. Rain
6. Wind
7. Snow (snow load)
8. Frost
9. Hail

Point 1 to 4 are mostly depending on the system while the other points are given by the local climate. Except point 9. hail, all operational and of course all weather conditions are included in the program during the one year period. The snow and wind load test is of minor importance for glass covered collectors; for the plastic covers, some additional research work is planed in the near future and is not tested in this procedure. Special attention is paid to the influence of high temperatures and rain.
To control the actual condition of each collector, certain parameters are monitored during the test:
    - performance (efficiency and thermal heat loss coefficient)

- stagnation parameter (representing the stagnation temperature)
- daily visual inspections (including a report on every failure, damage, discoloration, condensation, outgassing products etc.)

To determine the quality level and to decide if the collector is capable to achieve the minimal lifetime of 15 years, a score system was developed. The idea is quite simple: at the beginning each collector has the maximal score. For the decrease of the efficiency or the increase of the thermal heat loss coefficient, a certain deduction from the maximum score is calculated. The same procedure is applied for any damages or deterioration of materials during the test period. The deduction is calculated by comparing the costs for the repair with the price of the new collector. To determine, if the collector will survive the minimal lifetime of 15 years without significant reduction in performance and without major defects, the resulting score should be above a specified score limit.

3. Test method

3.1 The single test steps

The test program described in chapter 3.2 is divided in the following test steps:

- Stagnation test: no flow through the collectors
- Internal shock test: after heating up the collector to a specified temperature level, heat transfer fluid at ambient temperature is injected directly into the collectors
- External shock test: after heating up the collectors to a specified temperature level, the rain equipment starts working. Each collector is equipped with a single rain distributing system; if the ambient temperature drops below $0^{\circ}C$, the rain system stops. The amount of water is about 5 litres/min.$*m^2$ and is distributed over the surface

- Rain penetration test: the rain equipment starts
  working similar to the external shock test.
- Temperature cycle test: variation in temperature according to given temperature profiles. The minimum inlet
  temperature is at ambient temperature; the maximum temperature should not exceed the stagnation temperature of
  the tested collector.
- Determination of the efficiency and the thermal heat loss
  coefficient in accordance with the Swiss standard
  SN 165001/1.
- Pressure test: The operational pressure of the test loop
  is set to a maximum pressure of 4 bars. In the closed
  loop solar systems used in Switzerland, the normal
  pressure is never above 3 bars; the installations are
  protected by a relief valve set at 3bars.
- Frost test: During winter time of the one year test
  period, the collectors will automatically be exposed to
  frost conditions.

3.2 Test program

The test steps described in chapter 3.1 are combined to a test
program which is suited to the Swiss climate. The computer controlled test cycle lasts 48 hours, i.e. 2 whole days (see chapter 3.3).
Operating time of the single test steps (in hours):

- Normal operating condition in the temperature range   17
  from   20°C to 80°C (determination of the collec-
  tor efficiency and determination of thermal heat
  loss coefficient during the night)
- Stagnation condition (no flow through the collector)   15
- Temperature cycles   11
- External shock   2
- Internal shock   4
- Rain penetration test   4

Remark:

Some tests are carried out simultaneously during an other test phase. Thus the sum of the duration of each test listed above is greater than 48 hours.

3.3 Test cycle

The flow of the test cycle is independent of the weather condition. Even if there is no sunshine the mean temperature of the collectors during the efficiency measurement is kept at the required temperature level above the ambient temperature.
The temperature cycles, the external and internal shocks and the rain penetration tests are carried out between late afternoon and midnight. After midnight until 5 A.M., the thermal heat loss coefficient is determined. During the following few hours, there is no test to ensure that the collectors can cool down near the ambient temperature. Therefore condensation on the inside of the cover could be formed.

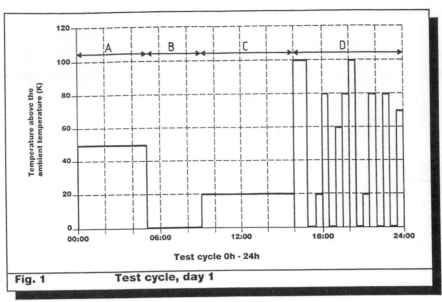

Fig. 1   Test cycle, day 1

A: Determination of the thermal heat loss coefficient   B: Stagnation (no test)
C: Determination of the efficiency   D: Changing load condition

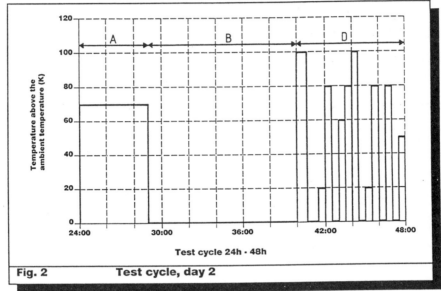

Fig. 2   Test cycle, day 2

A: Determination of the thermal heat loss coefficient   B: Stagnation   D: Changing load condition

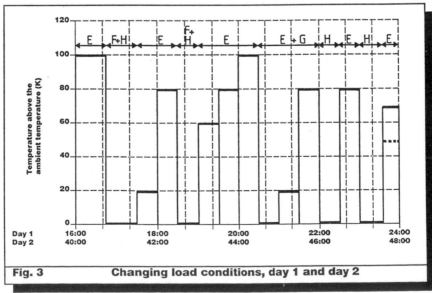

Fig. 3   Changing load conditions, day 1 and day 2

E: Temperature cycle test   F: External shock   G: Rain penetration test   H: Internal shock

126.

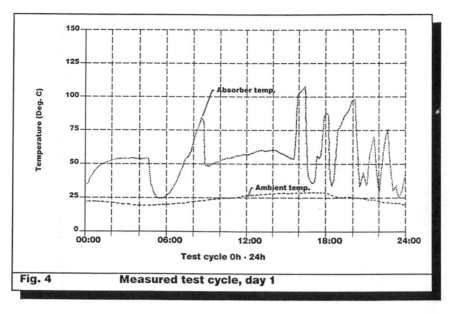

Fig. 4   **Measured test cycle, day 1**

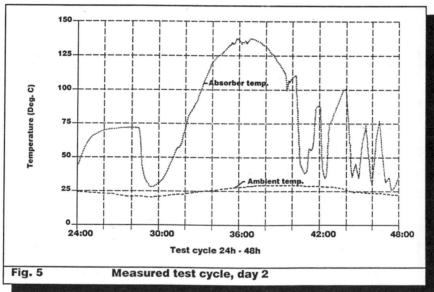

Fig. 5   **Measured test cycle, day 2**

3.4 Test facility

Figure 6 shows the scheme of the outdoor test facility at the "School of Engineering in Rapperswil" (Switzerland). The facility consists of 3 twin test frames, each test frame together with the heating and cooling unit forms one module. If there is the demand for more test places, several modules can be added.

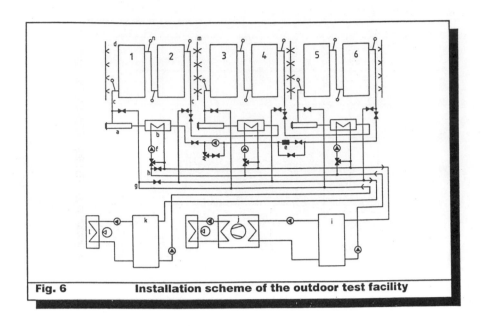

| Fig. 6 | Installation scheme of the outdoor test facility |

3.5 Instrumentation, data acquisition, visual inspection

Instrumentation:

Temperatures: o inlet and outlet temperature of each collector
o absorber temperature of each collector
o ambient temperature

The accuracy of the calibrated Pt 100 sensors is better than +/-0.05°C. After each one year test period the temperature sensors are recalibrated in a thermostatic bath.

Flow rate: o the flow rate of the collector loop is measured by an inductive flow meter, the accuracy is +/- 0.5% of the measured value.

Irradiance: o global irradiance in the collector plane
o diffuse irradiance in the collector plane

The accuracy is according to WMO classification for a 1st class pyranometer.

Wind velocity: o measured horizontal in front of the test
facility

The accuracy is +/- 0.5m/s

3.6 Data acquisition

Data are taken every 30 seconds. Measurement intervals vary
from 2 minutes up to 15 minutes, depending on the test steps.
An average value is calculated by averaging the collected data
within a measurement intervale.

3.7 Visual inspection

Daily at 9 A.M., exept saturday and sunday the collectors are
visually inspected.
This inspection belongs to the most important information to
qualify a collector.
Close attention is paid to fittings, joints, outgassing
products, corrosion, color-faiding and other details.
A concise report is filled daily and photographs are taken if
major changes occur.

4. Data evaluation

4.1 Frequency diagram

To quantify the load conditions during the test and to compare
two different test periods in terms of the temperature load, the
measured absorber temperatures are classified in $10^{o}C$ bandwidth.
The results are presented in a frequency diagram; i.e. number of
hours, where the absorber temperature lies within a given temperature band (i.g. from $70^{o}C$ to $80^{o}C$).
In the same way the time of operation are classified according
to the global irradiance received in the collectors' plane. The
bandwidth of the global irradiance is $100W/m^{2}$.

4.2 Determination of the collector parameters

The collector parameters are determined according to the Swiss Standard SN 165001/1, the following parameters are determined:

- optical efficiency for direct radiation perpendicular to the collector plane A_0
- optical efficiency for diffuse radiation A_{diff}
- thermal heat loss coefficient at small temperature differences between the collector and the environment K_0
- thermal heat loss coefficient, measured during night time K_{0N}

5.   Qualification

5.1 Requirements

- Collector installation, flow rate and pressure are according to the manufacturers' specifications
- The minimum tilt angle for ventilated flat plate collectors is 30°
- In function, the collector temperature never drops below the ambient temperature
- No hail with grains larger than 30mm in diameter
- No unnatural impacts

5.2 Restriction

- Internal corrosion, due to the use of non suitable heat transfer fluid, is not considered
- Major changes in design or construction of a collector by the manufacturer can strongly influence the long term behaviour

5.3 Score system

Lifetime and quality shall be expressed by a score system. This system is still refined.
Yet, the comparison between test results and the results of in

use conditions exposure has not been completed. The work is being continued.

The maximum score is 3000 points and represents the highest possible quality level. Deductions are depending on:

- damages and material deterioration
- decrease in efficiency
- increase of thermal heat loss coefficient

5.4   Score calculation

5.4.1 Rating of damages and material deterioration

A. Damages and material deterioration occurred during the test period

B. Damages and material deterioration detected after the dismantling of the test collectors

A. Damages occurred during the test period:

All damages occurred during the test period are reported in the daily inspections. Possible damages are:

- broken cover
- untight joints or fittings
- other visible damages

o Total loss:

The repair costs of the damages are more expensive than the price for a new collector. Any further discussion about lifetime or quality level is meaningless.

o Reparable damages or changes:

The repair costs are lower then the price for a new collector. The repair costs are determined by the manufacturer. The price covers parts and work.

Score reduction:

$$R_s = \frac{P_s * 1000}{P_n}$$

R_s = Deduction of the score in terms of damages
P_n = Price for the new collector
P_s = Accumulated repair costs during the test period

B. Damages detected after the dismantling of the collector

There are two major problems witch will lead to a high humidity inside of the collector.

        - untightness against rain
        - poor or no ventilation

Dependent on the general construction and the selected materials, high humidity or wetness can lead to serious corrosion problems. Corrosion affecting the absorber (e.g. aluminium rollbond) or the casing (e.g. thin galvanized sheet metal) can destroy a collector in a short period of time.

Judgement of the observed damages in the collector:

- Local corrosion attack on
  the absorber which leads almost
  certainly to untightness   =>   total loss

- Local corrosion attack on
  the casing (more than 20 spots
  per m^2)   =>   total loss

- Uniform surface corrosion is
  usually of minor importance.   =>   no point reduction

If there are any uncertainties about corrosion attacks a corro-

sion specialist is consulted for the decision whether a corrosion attack is of importance for the achievement of the required
lifetime of a collector or not.

- Small leaks caused by detachment
  of the sealer or any other
  small leaks   => reparable damage

The rating procedure is similar to the one described in chapter
5.4.1 A.

5.4.2 Rating of the changes of the optical efficiency

If the optical efficiency of the collector decreases during the
test period a score reduction is calculated as follows:

Score reduction:

$$R_{Ao} = \frac{(A_{oa} - A_{oe})}{A_{oa}} * 3000$$

R_{Ao} = Deduction of the score due to decrease of the optical
efficiency

A_{oa} = Mean value of the optical efficiency of the first month of
the test

A_{oe} = Mean value of the optical efficiency of the last month of
the test

Note: If the first or the last month don't allow to determine the
optical efficiency because of bad weather the next possible
value is used.

5.4.3 Rating of the change of the thermal heat loss coefficient

If the thermal heat loss coefficient of the collector increases
during the test period a score reduction is calculated as
follows:

Score reduction:

$$R_{KoN} = \frac{(K_{oNe} - K_{oNa})}{K_{oNa}} * 1000$$

R_{KoN} = Deduction of the score due to increase of the thermal heat loss coefficient

K_{oNa} = Mean value of the thermal heat loss coefficient of the first month of the test

K_{oNe} = Mean value of the Thermal heat loss coefficient of the last month of the test

5.4.4 Final score

$$P_Q = P_{max} - R_s - R_{Ao} - R_{Ko}$$

P_Q = Resulting final score

P_{max} = Start score; P_{max} = 3000 points

5.5 Quality classification

excellent	:	$2800 < P_Q \le 3000$
good	:	$2400 < P_Q \le 2800$
sufficient	:	$2000 < P_Q \le 2400$
insufficient	:	$P_Q \le 2000$

Assessment of the lifetime based on the resulting score

A collector can survive a minimal lifetime of 15 years, if each of the two collectors scores higher than 2000 points. It is assumed that the pertinent boundary conditions (chapter 5.1, 5.2) have been applied.

6. Results

Remark:

The here presented results are the current state of the ongoing research work on the lifetime assessment of whole collector

elements. With the further development of the test procedure the
results can differ to the here presented ones; especially the
comparison of the test results with damages or changes after
real in use conditions might influence the proposed test procedure.

Three different brands of flat plat collectors were tested from
December 1986 until December 1987. The results are discussed
according chapter 5.
From each product 2 collectors were simultaneously tested

Description of the test collectors:

Product A: single glazed (glass), selective absorber coating
             (black chrome on nickel on copper), mineral wool
             insulation, galvanized sheet steel casing, price
             $ 180.--/m^2

Product B: double glazed (plastic), selective absorber coating
             (paint on copper), textile wool insulation with
             asphalt exterior shell as casing, price $ 120.--/m^2

Product C: single glazed (glass), selective absorber coating
             (black chrome on stainless steel), mineral wool
             insulation, painted sheet steel casing, price
             $ 380.--/m^2

Results of product A (Test collector 1 and 2):

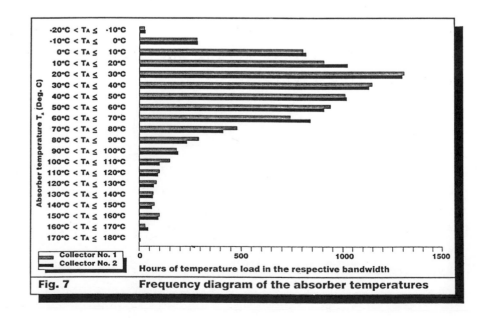

Fig. 7   Frequency diagram of the absorber temperatures

135.

The frequency diagram shows the temperature load during the test
period.
No damages were observed during the test period but two significant changes were noted. First little condensation on the
bottom of both collectors was formed. Secondly, after 6 month of
operation corrosion started on the inside of the frame.
After the dismantling of the collectors many serious local
corrosion attacks on the inner side of the galvanized sheet
metal casing was detected. Also parts of the frame showed many
local corrosion attacks. One reason for the strong corrosion is
the missing ventilation. The thin sheet metal would be perfo-
rated after few years of operation under Swiss climate conditions.

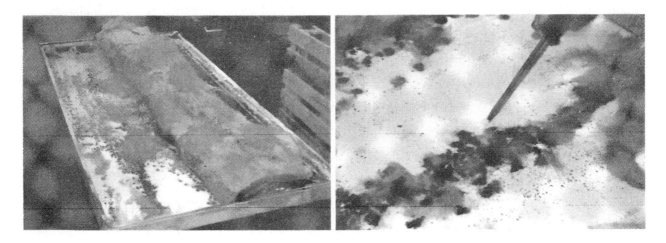

Fig. 8   Local corrosion attacks collector 1 and 2

According the definition in chapter 5 the collector would be
considered a total loss. Nevertheless the optical efficiency and
the thermal heat loss coefficient during the test phase is
displayed in the following figures:

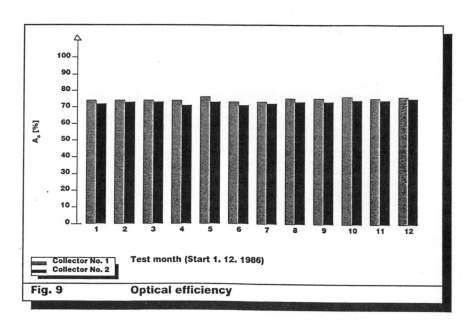

Fig. 9   **Optical efficiency**

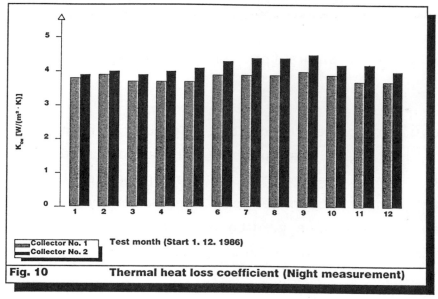

Fig. 10   **Thermal heat loss coefficient (Night measurement)**

The optical efficiency and thermal heat loss coefficient
remained stable during the entire test.

Final score

Collector 1: P_{Q1} = 3000 − 1000 − 0 − 0 = 2000 points

Collector 2: P_{Q2} = 3000 − 1000 − 0 − 26 = 1974 points

The resulting score is not above 2000 points, thus the quality
level is insufficient and this product will not survive the
required minimal lifetime of 15 years.

Results of product B (Test collector 3 and 4):

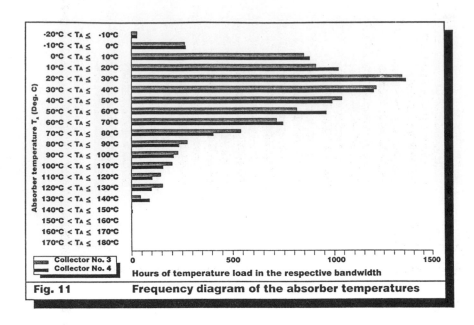

Fig. 11 **Frequency diagram of the absorber temperatures**

The frequency diagram shows the temperature load during the test.

No damages could be observed during the test but one change was observed. The initially transparent polycarbonate double plate cover, faded at the lower part to a more greenish colour. The often observed condensation between the two polycarbonate cover plates probably led to the growth of algue.
The dismantling of the collectors showed no serious changes.

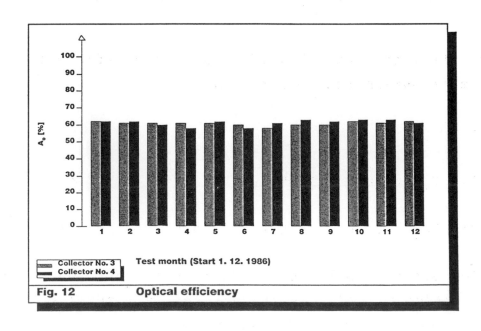

Fig. 12 **Optical efficiency**

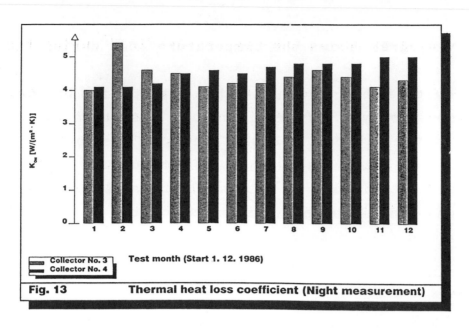

Fig. 13 **Thermal heat loss coefficient (Night measurement)**

The optical efficiency remained stable during the entire test.
The thermal heat loss coefficient of the collectors increased
slightly from 4 up to 4.3W/(m^2*K), and from 4.1 up to 5W/(m^2*K).

Final score

Collector 3: P_{Q1} = 3000 - 0 - 0 - 75 = 2825 points

Collector 4: P_{Q2} = 3000 - 0 - 48 - 219 = 2706 points

The resulting score is for both collectors above 2000 points,
the quality level is good and this product will survive the
required minimal lifetime of 15 years.

Results of product C (Test collector 5 and 6):

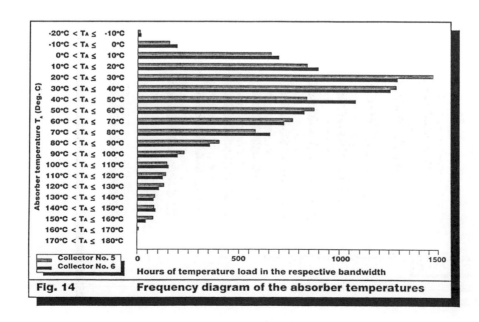

Fig. 14 **Frequency diagram of the absorber temperatures**

139.

The frequency diagram shows the temperature load during the test period.

No damages could be observed during the test period but one change has been observed. After 7 month of operation little condensation on the bottom of both collectors was formed. Dismantling of the collectors showed a uniform surface corrosion attack on the inlet fitting of both collectors. Also small leaks on the foreseen additional possibilities for the in or outlet of the collector could be observed.

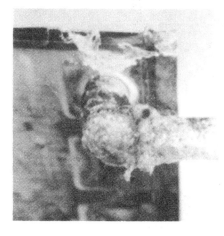

Fig. 15 Corrosion attack on the inlet fittings of collector 5 and 6

Fig. 16 Small leaks collector 5 and 6

These small leaks are considered reparable damages and can be sealed. The score deduction is calculated according to chapter 5.4.1. The cost for this repair is estimated at $90.--. The collector is well ventilated. Penetrating humidity will dry out.

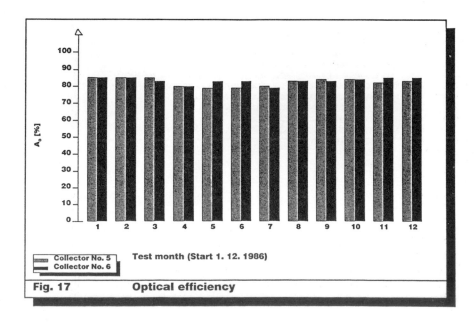

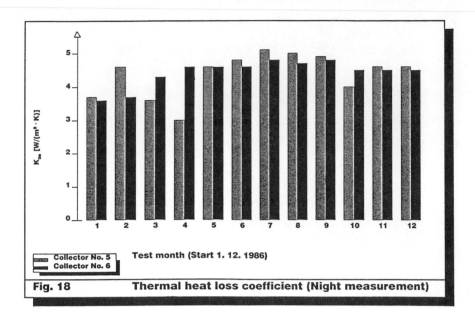

Fig. 18 Thermal heat loss coefficient (Night measurement)

The optical efficiency remained stable during the whole test.
The thermal heat loss coefficient increased slightly from 3.7 up
to 4.6W/(m^2*K) and from 3.6 up to 4.5W/(m^2*K).

Final score

Collector 5: P_{Q1} = 3000 - 237 - 71 - 243 = 2449 points

Collector 6: P_{Q2} = 3000 - 237 - 0 - 250 = 2513 points

The resulting score is for both collectors above 2000 points,
the quality level is good and the collector will survive the
required minimal lifetime of 15 years.

7. Conclusions

By means of this test procedure weak points of flat plate solar
collectors exposed to Swiss climate, can be found.
The minimal test period of 1 year can not be shorter because the
determination of the collector parameters is seasonally dependent and most of the observed effects can only be detected after
several months of testing.
The score system has to be further refined by means of comparison with results from collectors exposed to real conditions.

141.

ERRORS IN INTEGRATING SPHERE MEASUREMENTS DUE TO SPHERE GEOMETRY AND SAMPLE TEXTURE.

A. Roos, C-G. Ribbing and M. Bergkvist
Inst. of Technology, Uppsala University
Box 534, S-751 21 Uppsala, Sweden

ABSTRACT

It is experimentally demonstrated that for a class of samples with oriented surface texture, significant errors can arise in total and diffuse reflectance measurements with an integrating sphere. For strongly oriented polishing grooves on copper the error can be as large as 30%, if the grooves are oriented so as to give maximum entrance port losses. The same kind of error, albeit smaller, appear e.g. when rolled metal surfaces are studied. When monitoring the reflectance of metal based solar absorber surfaces the sample must therefore be properly oriented on the sphere port.

A mathematical model is proposed which takes account of entrance port losses and the fact that a fraction of the scattered light is not diffused in the sphere until after the reflection against the coating opposite to the sample. The two parameters to describe these anomalies can be determined in a straightforward experimental procedure and it is demonstrated that the corrections are significant for highly reflecting samples. The influence on the evaluation of solar optical properties, however, is small.

INTRODUCTION

Most standard spectrophotometers can be equipped with an integrating sphere. This makes it possible to measure the reflectance and transmittance of samples which scatter the incident radiation in a nonspecular manner. The scattered radiation is collected by the integrating sphere, and a detector monitors the average light intensity within the sphere. The theory of the radiation balance which is set up has been treated in detail by several authors[1-4]. To obtain a correct reflectance or transmittance value, the sphere wall has to be exactly spherical and coated with a highly reflecting, perfectly diffusing (Lambertian) material. It is also essential that no light leaving the sample surface can reach the detector directly without having been reflected by the sphere wall. Many different designs of integrating spheres have been reported in the literature, each with its special feature and often with a special application in mind. Possible errors arising from various geometrical imperfections have been summarized by Clarke and Compton[5].

Commercially available integrating spheres are usually double beam instruments that have been designed to operate in both transmittance and reflectance modes. In figure 1 the outline of the Beckman 198851 sphere is shown. This is the sphere used in this work. It can be seen that a fairly large proportion of the inside area is occupied by various ports. The operating instructions provided are not always complete and the integrating sphere theories found in the literature are often depressingly complicated for the user who simply wants to know the reflectance of a sample. In this paper we give an example of a surface geometry that will lead to significant errors unless special care is taken[6]. We also propose a simple formalism that will give a correct interpretation of the sphere signal output for most samples. The importance of correct sample orientation and signal interpretation for the evaluation of the total solar absorption α of an absorber surface is demonstrated. Only the reflectance mode is considered, but the same general ideas can be applied in transmittance measurements.

SPHERE AND SAMPLE GEOMETRY

The signal output of the instrument is in the simplest approximation the ratio between the reflectance of the sample and the reference plate. It is really the ratio between the amount of light absorbed by the detector when the sample is illuminated and when the reference plate is illuminated. Only a small part of the reflected light will reach the detector, in general after having been reflected by the sphere wall many times. If the sphere wall is a perfect Lambertian reflector then the light is evenly distributed within the sphere after the first reflection against the sphere wall. The reference plate is usually identical with the sphere wall (TiO_2, $BaSO_4$ or PTFE) and will therefore directly result in a homogeneous distribution of the light within the sphere. The sample, on the other hand, is generally not an ideal Lambertian reflector and will produce an irregular distribution of the reflected light, thus giving rise to a signal which is not proportional to the sample reflectance. This is the essence of the problem we shall address in this report.

To demonstrate the effect a model sample was prepared in the following way: a copper plate was polished in one direction with soft steel wool. This resulted in approximately parallell fine grooves in the copper surface. In figure 2 the reflectance spectra for such a sample are shown with the orientation of the grooves as parameter. θ=0° corresponds to vertical alignment of the grooves and θ = 90° to horizontal alignment.

The total reflectance is the sum of the diffuse and the specular reflectance

$$R_{tot} = R_{spec} + R_{diff}$$ (1)

The diffuse signal is measured with the specular exit port (no 5 in fig. 1) open. It can be seen that the sample orientation has a drastic influence on the recorded total and diffuse reflectance spectra. In figure 3 the diffuse signal R_d is shown as a function of θ between 0 and 90° for two different samples. The two different wavelengths correspond to two different detectors, and in both cases the recorded signal varies between 0.45 and 0.70. This is obviously a source of a significant error and it is important to understand the mechanism behind this behaviour. The explanation is depicted in figure 4 which is a photograph of the interior of the sphere taken through the reference port while the sample is illuminated. The grooves in the sample act as a "continuous grating" and most of the reflected light is scattered in a plane perpendicular to the grooves. On the photograph it can be seen how the reflected light from the surface for θ = 60° illuminates a band across the sphere wall. The ports seen on the picture are the two entrance ports on each side of the closed specular exit port (cf. fig. 1). It is easy to imagine that a considerable part of the reflected light escapes through the two entrance ports when θ = 0°. This corresponds to the low limit of the reflectance values in figures 2 and 3. For a sample orientation in the interval 55° ≤ θ ≤ 125°, the reflected light disc is contained within the sphere and a correct reading should be expected. The signal in fig. 3 reaches a maximum near 90° orientation, though, and this is due to the design of the screens which protects the detector from a "direct hit" from the sample. What happens at this θ value is that when the light disc is near the detector screens, part of this light falls on the <u>inside</u> of the rear screen and hence illuminates the detector indirectly in an irregular way. The resulting signal is therefore too high. The situation can be summarized by saying that the sample must be positioned on the sample port in such a way that as much as possible of the reflected light is contained within the sphere without striking the immediate vicinity of the detectors. Fig. 3 tells us that $\theta \sim 60°$ is the best choice in the visible, while $\theta \sim 110°$ is the best choice in the near infrared for this type of sample.

In figure 5 we show how sample misalignment can affect the result when solar absorber surfaces are being studied. The error in the total solar absorption value α can be of the order of several percent because of this effect only.

MATHEMATICAL MODEL

The signal levels in figures 2 and 3 are the "raw" signals from the spectro-photometer. Even when the proper sample alignment has been established, these signals have to be interpreted in a correct way to give the reflectance value of the sample. It is quite clear that the reflected intensity distribution from our copper sample differs from the Lambertian distribution of the reference plate. Going back to figure 4 we can see how the specular signal is established. When the circular port is opened the light that in figure 4 falls on that port escapes and does not contribute to the detector reading. This leaves the diffusely reflected light in the sphere.

A basic mistake is to treat the resulting diffuse signal in the same way as the diffuse signal from the reference plate. From figure 4 it is obvious that a large part of the scattered light is diffused in the sphere in exactly the same way as the <u>specular</u> signal, i.e. by the $BaSO_4$ coating opposite the sample. This means that the total reflectance should be written

$$R_t = R_s + R_d = R_s + R_{dd} + R_{sd} \tag{2}$$

where R_s is the specular reflectance, R_{dd} is the true diffuse component and R_{sd} is the "specularly diffuse" component.

Let us consider the spectrophotometer readings in the different situations. The signal output for the reference reading when a $BaSO_4$ plate is placed on both the sample and reference port is

$$S_1 = \frac{F R_B}{F R_B} A = A \tag{3}$$

where R_B is the reflectance of the $BaSO_4$-plates, A is the spectrophotometer amplification factor and F is the fraction of the reflected radiation that is contained within the sphere, i.e. the fraction 1-F is lost through the entrance ports. Note that, owing to optical imperfections, A is in general not independent of wavelength. Eq. (3) is strictly correct only if the two $BaSO_4$ plates are identical, which is assumed in the following. If R_B is known, a reference signal is obtained from the ratio

$$S_R = S_1/R_B = A/R_B \tag{4}$$

When the $BaSO_4$-plate on the sample port is replaced by a sample, the signal S_2 is obtained:

$$S_2 = \frac{R_s R_B + F R_{dd} + R_{sd} R_B}{F R_B} A \tag{5}$$

R_s and R_{sd} are both multiplied by R_B because of the extra reflection against the sphere wall these components must undergo before they are completely diffused. If we assume that the fraction B of the scattered light is Lambertian and the fraction 1-B is "specular", eq. (5) can be rewritten

$$S_2 = \frac{R_s R_B + FBR_d + (1-B)R_d R_B}{F R_B} A \tag{6}$$

If the specular exit port is opened the signal becomes

$$S_3 = \frac{FBR_d + (1-B)R_d R_B}{F R_B} A \tag{7}$$

since the component R_s escapes. R_d and R_s can now be expressed as functions of the three spectrophotometer signals S_1, S_2 and S_3, the two parameters F and B and the known reflectance R_B.

$$R_d = \frac{F}{FB + (1-B)R_B} \frac{R_B S_3}{S_1} \tag{8}$$

$$R_s = (S_2 - S_3) \frac{F}{S_1} \tag{9}$$

The parameter F can in principle be estimated from the ratio of the entrance port area to the total sphere area. If the intensity distribution of the reflected light from the $BaSO_4$-plate differs from the ideal Lambertian, this ratio may not coincide with the fraction of reflected light escaping through these ports. It is then better to determine the factor F experimentally. In figure 6 we show how this was done using a specular evaporated aluminium mirror. The diffuse signal S_3 for this mirror is zero and by comparing the sphere signal to the absolute reflectance value obtained with a V-W attachment F was found to be 0.98. This is in fact smaller than the surface area ratio of 0.99 which indicates that the $BaSO_4$ scatters the light more in the forward direction than the ideal Lambertian surface.

In figure 7 the ideas of the mathematical model are illustrated. The reflectance curves for an ordinary rolled aluminium sample are shown for different values of the parameter B. Also shown for comparison is the incorrect value S_2/S_B which is given in the operating instructions for the sphere. The curve for B=1 corresponds to the assumption that all the scattered light is Lambertian. This is the most widely adopted approach, but it can be seen that in the near infrared, the reflectance curve exhibits spurious dips that can be recognized as originating from the $BaSO_4$ coating. This structure nearly vanishes for B = 0.3 and is reversed when B approaches zero. From basic principles it is difficult to determine the value of the parameter B, but the results presented in figure 7 show that B can be taken as the value,for which the spurious $BaSO_4$-like structure in the reflectance curve is minimized. We can also see that neglecting the subdivision of R_d into two components, as in eq. (2), can result in an error of the order of a few percent over the whole spectrum and as high as ten percent in the near infrared.

Taking these considerations into account for solar absorber surfaces can in some cases be important, but in general the variations in the α-value due to different values of the parameter B are very small. The lower the reflectance is, the smaller is the variation in α. For the samples in fig. 5 the variation in α is of the order of a few tenths of a percent.

SUMMARY

We have pointed out the importance of proper sample alignment when using an integrating sphere and shown that for some structured samples the error can be as high as 30 % if the sample is incorrectly positioned. We have also shown that turning a steel wool polished sample 180° in small steps on the sample port provides important information of the sphere geometry and helps in understanding the signal output from the instrument. Neglecting these considerations for the evaluation of the total solar energy absorption can lead to significant errors in α. We also present a mathematical model which improves the analysis for most samples with a non-homogeneous distribution of the scattered light. The proposed model entails dividing the diffusely reflected light into two components, which have to be analyzed in different ways. This distinction is generally valid for any non-Lambertian surface and is not a peculiar feature of the polished surface used in this work. The detailed results presented in this paper are specific for the Beckman integrating sphere used at our laboratory, but the ideas are generally valid and easy to apply to other integrating sphere geometries.

ACKNOWLEDGEMENT

This work has been sponsored by the Swedish Council for Building Research.

REFERENCES

1. J.A.J. Jacquez and H.F. Kuppenheim "Theory of the Integrating Sphere", J. Opt. Soc. Am. 45, 460 (1955).

2. B.J. Hisdal "Reflectance of Perfect Diffuse and Specular Samples in the Integrating Sphere", J. Opt. Soc. Am. 55, 1122 (1965).

3. B.J. Hisdal "Reflectance of Nonperfect Surfaces in the Integrating Sphere", J. Opt. Soc. Am. 55, 1255 (1965).

4. D.G. Goebel "Generalized integrating sphere theory", Appl. Optics 6, 125 (1967).

5. F.J.J. Clarke and J.A. Compton "Correction Methods for Integrating-Sphere Measurement of Hemispherical Reflectance", COLOR Res. and appl. 11, 253 (1986).

6. A. Roos, C-G. Ribbing and M. Bergkvist "Anomalies in Integrating Sphere Measurements on Structured Samples", proceedings "Workshop on Optical Property Measurements Techniques", October 1988, ISPRA, Italy.

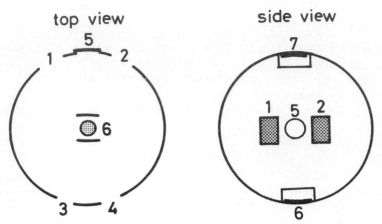

Fig. 1. Integrating sphere design with components as follows: 1) entrance port, reference beam 2) entrance port, sample beam 3) sample port 4)reference port 5) exit port, specularly reflected beam 6) lead sulphide detector 7) photomultiplier detector.

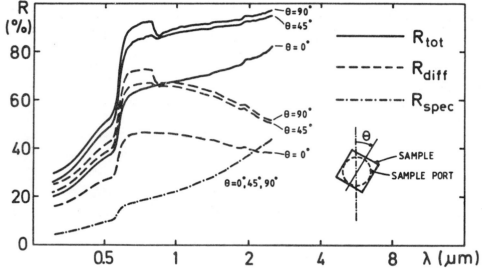

Fig. 2. Reflectance spectra for a polished copper plate.

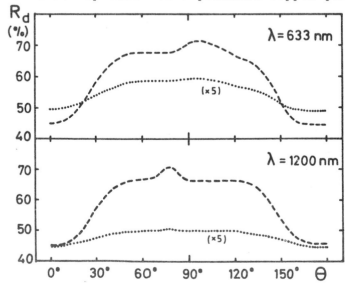

Fig. 3. Diffuse reflectance at two fixed wavelengths vs sample orientation as defined in fig. 2. ---- Steel wool polished copper plate ···· oxidized copper plate.

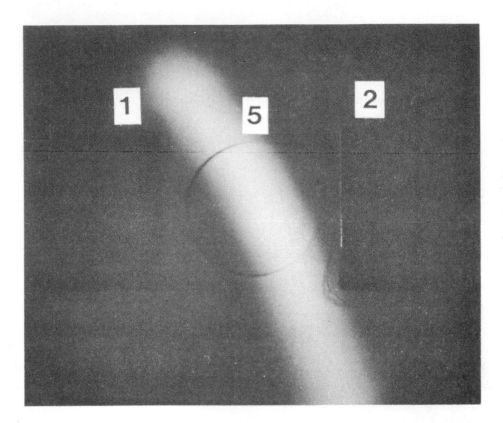

Fig. 4. Photograph showing the inside of the sphere illuminated by the reflected light from a steel wool polished copper surface. The numbers refer to fig. 1.

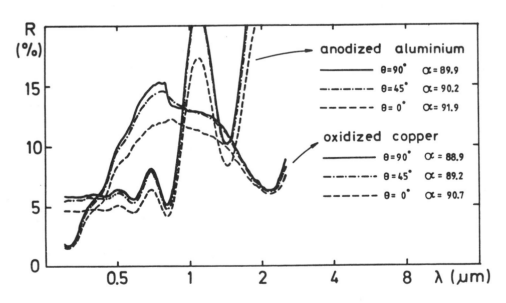

Fig. 5. Total reflectance spectra for two solar absorber surfaces with sample orientation θ as indicated.

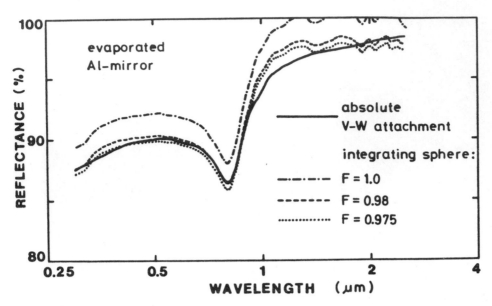

Fig. 6. Spectral reflectance for an evaporated AL-mirror as measured with a V-W attachment and the integrating sphere. The sphere value is from eq. 9.

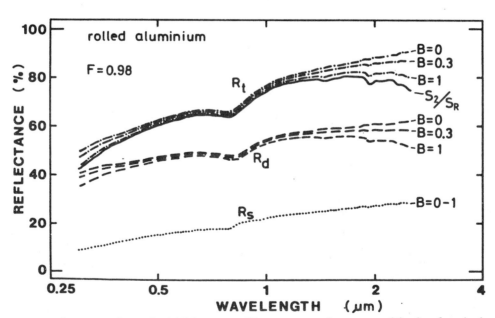

Fig. 7. Spectral, total and diffuse reflectance for a rolled aluminium surface as calculated from equations 8 and 9 using parameters F and B as indicated.

SPECTRAL MEASUREMENT OF THE INFRARED REFLECTANCE

K. Gindele and M. Köhl

University of Stuttgart
Institute for Theory of Electrical Engineering
Pfaffenwaldring 47, D-7000 Stuttgart 80, F.R.G.

INTRODUCTION

The measurement of the directional/hemispherical reflectance of diffusely reflecting samples requires the determination of the entire flux reflected from a directionally irradiated surface. This could be done by mathematical integration of the spectral bidirectional reflectance, but the measurement of the bi-directional reflectance /1/ is a difficult, costly, and time-consuming undertaking, mostly restricted to some individual wavelengths, and yields more information, namely the indicatrix, than is needed in most cases. If the knowledge of the indicatrix is not needed, it would be better to use a measuring method with a physical integrator, i. e. a device, which collects the hemispherically reflected radiation with no regard for their arbitrary angular distribution.

Such a physical integration can be achieved either by regularly reflecting mirrors of elliptic /2/, parabolic /3/ or spherical /4/ shape, or by integrating spheres with a diffusely reflecting coating. Both methods of hemispherical collection have their own strengths and weaknesses and for both the uniformity of the angular response is the point. This paper will focus on the use of integrating spheres for measuring the near-normal/hemispherical reflectance in the infrared spectral range.

MEASURING DEVICE

Integrating spheres for use in the IR are normally attached to Fourier transform spectrometers to reap the benefits of Fourier transform spectroscopy for overcoming the throughput penalty of integrating spheres described below, because in the spectral range of thermal radiation only relatively weak sources with a continuous emission spectrum exist. Since, in contrast to mirror devices, at integrating sphere arrangements the reflected flux is not focused on the detector, but distributed over the whole area of the sphere leading to a tremendous flux reduction by a factor of the order of magnitude of 1000. Fortunately multireflections again enhance the irradiance of the wall by a factor of about 10, yielding an overall flux reduction of about 100 when comparing same areas of irradiance and detection.

A schematic drawing of the optical system built up at the University of Stuttgart for measurements of the spectral near-normal/hemispherical reflectance in the IR spectral range as well as of the spectral normal emittance at 373 K is shown in figure 1.

Other systems are described in the references 5 to 7, a historical sketch of carried out arrangements is given in reference no. 8.

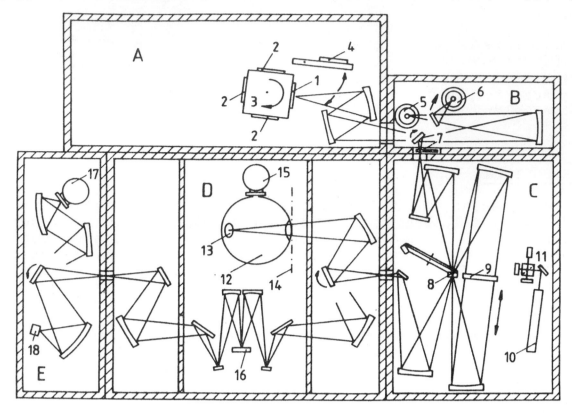

A: Emission accessories   D: Sample compartment
      1 Reference   (reflection accessories)
      2 Samples   12 Integrating sphere
      3 Heater   13 Sample port
      4 Gold mirror   (reference port below)
                                           14 Tilting axis
                                           15 Detector
B: Source chamber   16 Sample

C: Interferometer   E: Detector chamber

Figure 1: Schematic drawing of the optical system.

The integrating sphere is implemented in the sample compartment of an evacuable rapid-scan FTIR spectrometer, Bruker IFS 113v, which yields adequate experimental conditions for integrating sphere measurements in the wavelength range from 1.5 µm to about 15 µm. The following combination of spectrometer components is adequate for measurements in the NIR (usable range: 1.5 µm to 5.5 µm): tungsten lamp, Si/CaF_2 beamsplitter, and InSb detector. For measurements in the MIR (usable range: 2 µm to 14.3 µm) the combination globar source, Ge/KBr beamsplitter, and HgCdTe detector is used. The typical noise spectra as well as the corresponding spectral beam intensities for both configurations are shown in figures 2 and 3.

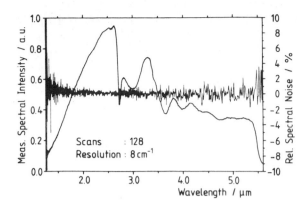 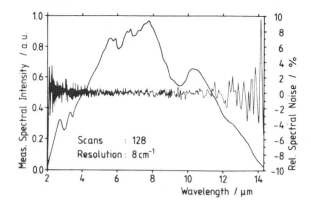

Figure 2: Single beam intensity and noise spectrum, configuration "NIR".

Figure 3: Single beam intensity and noise spectrum, configuration "MIR".

The instrument is also equipped with a specular reflectance accessory as well as with a turnable heating device for comparative radiometric measurements, which are described in some more detail in reference 9. In these two cases measurements can be taken between 1.5 µm and 25 µm by using a DTGS detector.

Diffuse reflection attachment

The integrating sphere, fabricated by Labsphere, has a diffusely reflecting gold coating and is designed for relative measurements with the comparison method. Both, sample and reference, are simultaneously attached to the sphere ports and act as parts of the sphere wall. In this arrangement no sphere error due to the different reflectances of the sample and of the reference occurs and the sample handling is very easy. Sample and reference can be irradiated alternately at an angle of 10° to the surface normal by tilting the whole reflection device. Specularly reflected radiation also hits the sphere wall and is included in the flux measurement. The detector measures the brightness of the sphere wall. It has a limited field of view so that it can detect neither the radiation directly coming from the sample or the reference nor from their specular reflexes. Details of the reflection unit are given in reference 10.

Sphere wall coating

The inner wall of the integrating sphere is coated with a diffuse gold coating. It consists of a 0.5 µm thick electro-deposit of gold over the "Lambertian" roughened aluminium substrate. The wavelength dependence of the diffusing properties of such a microrough surface is strongly dependent on the roughness parameters /11, 12/. Therefore the roughness has to be chosen carefully to provide the desirable properties over a wide spectral range. If the mean feature size is too small, the Lambertian behaviour of the surface deteriorates with increasing wavelengths as it is shown for sandblasted surfaces with an overcoat of gold in reference 11. This in fact appears for the surface of the inner wall of the sphere used up to now at the

University of Stuttgart (date of delivery: 1982) as can be seen
from measurements of the specular reflectance factor SRF /13/ of
a sample plate with the same coating (shown as straight line in
figure 4) and from the wavelength dependence of the angular
response of the sphere /14/. Nowadays Labsphere applies a more
suitable coating with a roughness average about 2.5 times larger
yielding higher reflectance /15/ and better diffusing properties
over a wider spectral range (specular reflectance factor SRF
shown as dashed line in figure 4).

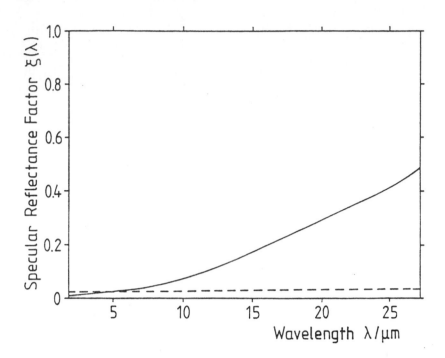

Figure 4: Spectral specular reflectance factor of the Labsphere
          diffuse gold coating for two different surface rough-
          nesses (measuring solid angle 0.125 sr).

The results of bidirectional reflectance distribution function
BRDF measurements performed on this surface at the University of
Arizona at a wavelength of 10.6 µm are reported by Willey /15/
for 10 and 30 degree incident angle and for an angle of incidence
of 20 degree by Hanssen and Snail /16/. These measurements of the
bidirectional reflectance show an almost Lambertian character to
the scattering at this wavelength. At the moment, the sphere in
our arrangement is replaced with a sphere wearing the new rougher
coating.

REFLECTION STANDARDS AND EVALUATION

The geometry in use for reflectance measurements is designed for
measuring in a relative manner: the signal of the sample is related to that of a reference. The problem of a suitable reference
is a key problem of measurements of the diffuse IR reflectance.
The requirements to the reference are:1. high reflection, 2. Lambertian reflection characteristic, 3. spectral neutrality, i. e.
without a spectral structure (as "grey" as possible), and last
not least: 4. calibrated reflectance. The accuracy of the measurement is directly depending on the achievement of these
properties.

Sulphur is one of the promising candidates for providing a suitable reference /17/, but it has not gained general acceptance, may be due to its sensitivity or its difficulty in handling or may be due to other reasons /18,19/. Looking for a stable, easy-to-use diffuse reflectance standard, today there only remain textured metal surfaces. Practically we know of two coatings. The one is the Labsphere diffuse gold coating described above, the other a flame sprayed aluminium coating on stainless steel, which can be got from National Physical Laboratory (NPL) in Teddington, U. K.. The first one satisfies the requirements 1 and 3 and, in its newer form, also number 2 for a sufficiently broad spectral range, but we could not get a surface calibrated in hemispherical reflectance. We also dont have any data about the homogeneity of the surface, and this helds for the second coating too. The NPL coating shows a weak structure in the spectral curve, has a relatively low reflectance (about 0.90) and we have no information about the stability. For an angle of incidence of 17.5 degree, there is almost no specular component, but the BRDF has not been measured.

As a conclusion: today there is no diffusely reflecting surface which really can confidently satisfy all requirements for a diffuse reflectance standard. But, there is some ongoing work at PTB and NPL to measure the near-normal/hemispherical reflectance of the Labsphere coating in its new version so that this coating will hopefully also achieve requirement no. 4 (calibrated reflectance).

At the University of Stuttgart we have used up to now a specularly reflecting gold mirror as reflection standard, because we saw the falsifying influence of the increase of the specular reflectance factor of the sphere wall with growing wavelengths and we wanted to be sure on the indicatrix of the reference /10/. This may be a good solution for the measurement of specularly reflecting samples - and many of the samples are going regularly reflecting in the IR /20/, but it is not a real solution of the task, since - for reasons which are elaborated in the next paragraph - for diffusely reflecting samples it only shifts the problem from the indicatrix of the reference to the reflectance of the sphere wall (which we dont know for sure). It would be the best in that case to use a diffusely reflecting reference and we are going to do so just when the calibrated reflectance data of the Labsphere coating will be provided to us.

For the evaluation of the measured data, we have to take into account that in sphere arrangements, in which the specular component is included in the measurement, the signal A_{sp} for a flux reflected specularly from a surface is not only proportional to the reflectance ρ of the surface, but moreover weighted by the reflectance of the sphere wall coating ρ_w

$$A_{sp} \sim \rho_w \, \rho,$$

while the signal A_{dif} for a flux reflected from a diffusely reflecting sample is

$$A_{dif} \sim \rho \, .$$

This result of a calculation /21/ easily can be made clear if we realize that the incident flux is reduced by the reflection at the sample by a factor ϱ without changing the spatial distribution. It hits the sphere wall in a beam and we haven't got the same situation as for a diffusely reflecting sample until the first reflection at the sphere wall has taken place.

Having that in mind we see a further problem of IR-active integrating spheres. The different weight of diffuse and specular component of the reflected flux, as mentioned above, does not cause severe troubles in the visible spectral range because there the reflectance of the wall is nearly one and moreover in this spectral range most samples reflect diffusely anyway. In contrast to that the reflectance of a diffuse gold coating in the IR is about 0.95 and many samples have a reflection character which changes from diffuse in the VIS to specular in the FIR. (This is particularly the case for those samples which have surfaces with a microroughness in the order of magnitude of 1 micron.) The effect of the different weighting of the components by integrating spheres in the IR is most important for samples which are highly reflecting in the infrared (e. g.: metals, selective solar absorbers).

As shown in reference 20, this error can be corrected arithmetically for a superposition of diffuse and specular reflection with a good accuracy, if the spectral specular reflectance factor is known from separate measurements of the specular or the diffuse component. Figure 5 demonstrates this evaluation method for a copper oxide layer.

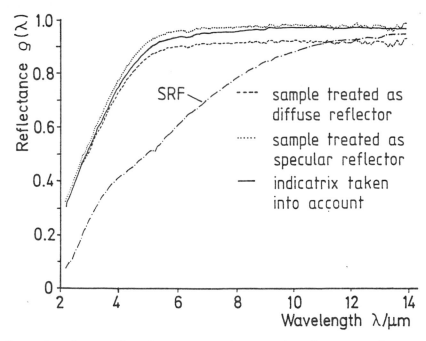

Figure 5: Spectral reflectance and spectral specular
          reflectance factor of a copper oxide layer.

The new sphere with the rougher coating is supposed to have a higher wall reflectance than the old one so that the integrating capacity of the sphere is only slightly distorted, thus it might come out that this correction could become unnecessary.

PROPERTIES OF THE MEASURING DEVICE

Angular response

The uniformity of the angular response of a integrating device (sphere or mirror) is very important for achieving a good accuracy of the reflectance measurements of diffusely reflecting samples. The angular response can be checked by measuring ruled or wedge-shaped highly reflecting samples. In our case two effects could be observed for the old sphere with the aid of such measurements.

1) A moderately enhanced response for radiation which hits the region of the sphere wall seen by the detector.
This result was anticipated for the applied sphere design which avoids baffles, but its effect on the measurement result of diffusely reflecting samples is small because the affected area is small compared with the total sphere area. Moreover the influence of this effect will be minimized by using a diffusely reflecting reference.

2) The angular response as a function of the wavelength is not uniform in all cases.
In those cases in which the geometry allows specular reflection in the direction of the detector we get spectral distributions rising with increasing wavelengths (according to figure 4) because the diffuse reflection is supported by regular reflection. In some other cases they slightly descend because the unavoidable loss of reflected radiation through the entrance port is amplified by regular reflection.

Both effecs are expected to have disappeared after replacing the old sphere by one with the new wall coating.

For samples with a "satin like" indicatrix (i. e. samples whose reflection behaviour is neither diffuse nor specular nor a superposition of both, but something in between, that means the indicatrix shows an enhanced reflection in a more or less wide angular range around the angle of regular reflection) a slightly reduced accuracy must be taken into account since the regular reflex, which hits the sphere wall in the direct neighbourhood of the entrance port, is "widened" as described above and an overproportional amount leaves the sphere through the entrance port (The area of the entrance port is 1.7% of the total area of the sphere surface).

Selfemitted radiation of the sample

In conventional spectrophotometers the distinction between measuring radiation and thermal radiation emitted by the sample is usually made by chopping techniques. This technique fails in FTIR spectroscopy, therefore the effect of this additional radiation depending on the emittance ε of the sample, which is for opaque samples coupled with the reflectance by the relation $\varepsilon = 1 - \rho$, has to be quantified.

In FTIR spectroscopy the sample, which in our case is part of an integrating sphere arrangement, is irradiated with "spectrally

coded" radiation, i.e. radiation having passed the interferometer. This dynamic signal would be well distinguishable from the
DC signal of the thermal radiation, but the selfemitted radiation
of the sample partly leaves the sphere and is partly redirected
to the sample after having passed the interferometer too. The
factor deciding on the error introduced by that superposition of
spectrally coded fluxes (Φ_{gl} coming from the globar source, and
Φ_{th} the thermal radiative flux from the sample which is at room
temperature) is the ratio Φ_{th}/Φ_{gl}. For the Stuttgart arrangement we have estimated this ratio to be smaller than 3×10^{-3}
(compare figure 6), thus the influence of the thermal radiation
can be neglected in our case. Please, notice in figure 6 the increasing effect of the sphere. The special wavelength dependence
of the curve "with sphere" is caused by the competition of two
effects, both governed by the reflectance of the sample: emission
of thermal radiation into the sphere, decreasing with increasing
reflectance, and reflection of part of this radiation into the
interferometer by the sample, proportional to the reflectance.

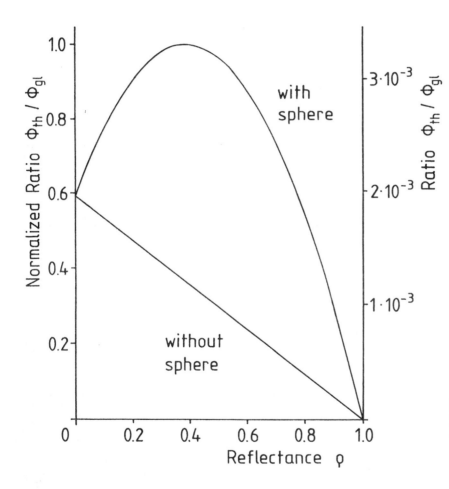

Figure 6:
Influence of the
thermal radiation
emitted by the
sample.

Accuracy

The photometric accuracy of the spectrometer (without sphere) is
0.1 %. Restrictions of the accuracy in reflectance terms are
mainly caused by two sources of errors: 1. uncertainty of the
spectral directional/hemispherical reflectance of the standard
surface used as reference in relative measurements and 2. non-
ideal properties of the sphere design and/or of the sphere wall
coating. The first source of uncertainty gives a systematic

error, while the second one yields an error the amount of which strongly depends on the indicatrix of the measured sample. At the moment, it is hardly possible to give a general statement on accuracy, but an idea is given by the result of a round robin measurement on highly reflecting diffuse test samples initiated by R.R.Willey /15/ (including measurements with 5 IR integrating sphere devices, a mirror integrator, and a heated cavity (Hohlraum)). The measured reflectance values differ in a range of about 0.1.

A comparison of the results of measurements taken on the NPL diffuse aluminium sample (see chapter "reflection standards and evaluation"), measured at NPL with the device described in reference 4, with those gained with the Stuttgart integrating sphere device showed agreement within 1 % of reflectance in the wavelength range from 5 µm to 12 µm, but higher deviations (up to 7 %) at both ends of the measuring spectral range (between 2 µm and 5 µm and between 12 µm and 14 µm) which are not understood up to now.

Comparison with other techniques

Figures 7 and 8 show for comparison the results of measurements on selective solar absorbers gained with the diffuse gold coated integrating sphere, a $BaSO_4$-coated sphere (for wavelengths up to 2.5 µm), and a specular reflection accessory (for wavelengths longer than 2.5 µm) as well as the radiometrically measured emittance. The good agreement between all curves is obvious for the specularly reflecting sample (figure 7) wich was deposited on a flat glass substrate. In figure 8, which gives the spectra measured on an aged copper oxide coating, which is a partly diffuse reflector, clearly show the deviation of the specular reflectance curve from the results gained with methods which take

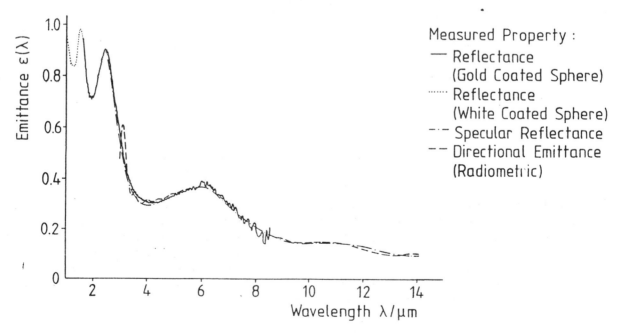

Figure 7: Spectral directional emittance of a graded Ni/MgF$_2$ cermet coating on glass, comparison of different measuring methods.

the diffuse component of the reflected flux into account, but these again are in good agreement between each other. In figure 8 the low wavelength limit of the radiometric measurement (sample temperature: 373 K) can be seen by the increase of noise in the spectral region below about 4 µm.

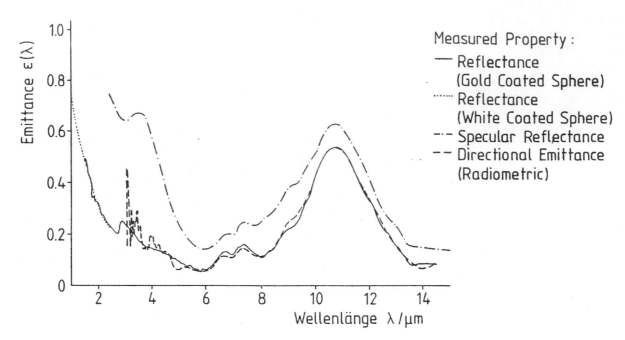

Figure 8: Spectral directional emittance of an aged copper oxide coating on copper, comparison of different measuring methods.

At the moment, the measurement of the hemispherical reflectance by means of an IR-active integrating sphere fails at about 15 µm due to the decrease of intensity at that wavelength. For a measurement at longer wavelengths mirror integrators have to be used or emission measurements on heated samples have to be taken which cause many more problems. These measurements make more demands on the sample form and/or substrate material and moreover the convenience of short measuring times and easy sample handling gets lost (compare also next chapter).

DISCUSSION

There are some advantages of integrating sphere arrangements attached to FTIR spectrophotometer systems, which make them best suited for investigations like ageing tests on absorbers (compare "Durability Testing and Service Lifetime Prediction of Solar Energy Materials" by Köhl, Gindele and Frei, contribution to this conference) in the course of which a lot of measurements has to be taken on samples which often have small curvatures either by cutting them out of real collectors or due to the heat treatment in accelerated tests. The described measurements dont take long measuring times and the sample handling is very easy, because a sphere geometry with wall mounted samples had been chosen. In contrast to most of the other possible arrangements small sample curvatures are allowed and there are almost no limitations in size and form of the spezimen if only the measured area is fairly plane and covers the measuring port. This allows to measure samples whose areas are large compared with the measuring area, a feature which is important for ageing investigations to avoid the influence of border effects on the measuring results. Additionally, there are no restrictions in the substrate materials.

The measuring temperature is room temperature, which is convenient and in most cases sufficient. There is no thermal load on the sample, neither by substrate heating (in radiometric devices) nor by strong irradiation (as is often the case when using mirror integrators). That is important for the measurement of temperature sensitive samples as e. g. semiconductors or some organic materials.

But, to our opinion integrating sphere devices are not very suited to measure the temperature dependence of the reflectance (remember the difficulties of distinction between measuring radiation and the selfemitted radiation), this would better be done with radiometric measurements.

CONCLUSIONS

Measuring arrangements with integrating spheres wearing a diffusely reflecting gold coating are well suited devices for the measurement of the directional/hemispherical reflectance and the number of installed sphere arrangements is growing. The usable spectral range reaches from 1.5 µm up to about 15 µm. The upper wavelength limit is not only caused by the descent of the signal intensity but also by the deterioration of the diffusing properties of the sphere wall.

There is ongoing work to improve these devices in wall coating, measuring geometry and radiation collecting to get a better angular response and/or a higher sensitivity. Despite the throughput penalty and the slightly reduced capacity of integration the use of integrating spheres in the IR has some important advantages, especially for the measurement of extensive series of samples. The measuring time is short, the sample handling is very easy, and material and size of the samples can be chosen almost freely. Even small curvatures of the measured sample area are allowed.

The angular response is supposed to be improved by the new rougher sphere wall coating. For a further enhancement of the measuring accuracy there is a strong need for a reflection standard with a Lambertian reflection character whose spectral reflectance is exactly known from calibration measurements.

ACKNOWLEDGEMENT

This work has been financially supported by the Minister for Research and Technology of the F.R.G (Contract No. 03E8020B).

REFERENCES

1. See e. g.: A. A. DeSilva and B. W. Jones,
    Bidirectional spectral reflectance and directional-hemi-
    spherical spectral reflectance of six materials used as
    absorbers of solar energy. Solar Energy Mater. **15** (1987),
    391-401.
    Compare also: B. W. Jones, A. A. DeSilva and J. W. Gannon,
    Goniometric measurements of reflectance at solar wavelengths
    and of emittance at thermal wavelengths. Contribution to
    this conference.
2. See e. g.: G. J. Dorman,
    The determination of radiative properties of surfaces using
    reflectance techniques. Report EUR 9520 EN, JRC Ispra 1984.
3. See e. g.: R. T. Neher and D. K. Edwards,
    Far infrared reflectometer for imperfectly diffuse speci-
    mens. Appl. Opt. **4** (1965), 775-780.
4. See e. g.: F. J. J. Clarke and J. A. Larkin,
    Measurement of total reflectance, transmittance and emissi-
    vity over the thermal IR spectrum.
    Infrared Phys. **25** (1985), 359-367.
5. R. R. Willey,
    Fourier transform infrared spectrophotometer for
    transmittance and diffuse reflectance measurements. Appl.
    Spectrosc. **30** (1976), 593-601.
6. W. Richter,
    Fourier transform reflectance spectrometry between 8000 cm^{-1}
    (1.25 μm) and 800 cm^{-1} (12.5 μm) using an integrating
    sphere. Appl. Spectrosc. **37** (1983), 32-38.
7. K. A. Snail, K. F. Carr,
    Optical design of an integrating sphere-Fourier transform
    spectrophotometer (FTS) emissometer. Proc. of SPIE Symp. on
    Infrared Optics, SPIE Vol. **643** (1986), 75-83.
8. K. Gindele and M. Köhl,
    Strengths and weaknesses of IR-active integrating spheres.
    Proc. of Workshop on Optical Property Measurement Tech-
    niques, Ispra Italy, 27-29 Oct, 1987, to be published.
9. K. Gindele, M. Köhl and M. Mast,
    Spectral directional emittance measurements in the wave-
    length range from 1 μm to 15 μm. Proc. of Fourier and Compu-
    terized Infrared Spectroscopy, SPIE Vol. **553** (1985),
    391-392.
10. K. Gindele, M. Köhl and M. Mast,
    Spectral reflectance measurements using an integrating
    sphere in the infrared. Appl. Opt. **24** (1985), 1757-1760.
11. R. A. Winn, D. P. DeWitt,
    Evaluation of a Fourier transform infrared spectrophotometer
    for measurement of diffuse reflectance. Proc. of 7th Symp.
    on Thermophysical Properties held at NBS Gaithersburg,
    Maryland, May 10-12, 1977, 285-294.
12. M. Köhl and K. Gindele,
    Determination of the characterizing parameters of rough
    surfaces for solar energy conversion. Solar Energy Mater.
    16 (1987), 167-187.
13. ASTM Standard E 429,
    Measurement and calculation of reflecting characteristics of
    metallic surfaces using integrating sphere instruments.

14. K. Gindele and M. Köhl,
    Measurement of near-normal/hemispherical reflectance and
    directional emittance in the mid-infrared. Proc. of SPIE
    Symp. on Passive Infrared Systems and Technology,
    SPIE Vol. **807** (1987), 160-164.
15. R. R. Willey,
    Results of a round robin measurement of spectral emittance
    in the mid-infrared. Proc. of SPIE Symp. on Passive Infrared
    Systems and Technology, SPIE Vol. **807** (1987), 140-147.
16. L. M. Hanssen and K. A. Snail,
    Infrared diffuse reflectometer for spectral, angular and
    temperature resolved measurements. Proc. of SPIE Symp. on
    Passive Infrared Systems and Technology,
    SPIE Vol. **807**(1987), 148-159.
17. J. T. Agnew and R. B. McQuistan,
    Experiments concerning infrared diffuse reflectance
    standards in the range 0.8 to 20.0 microns. J. Opt. Soc. Am.
    43 (1953), 999-1007.
18. M. Kronstein and R. J. Kraushaar,
    Sulfur as a standard of reflectance in the infrared.
    J. Opt. Soc. Am. **53** (1963), 458-465.
19. W. Erb und W. Richter,
    Reflexionswerte von gepreßtem Schwefelpulver im Spektral-
    bereich von 0,5 µm bis 5 µm. Optik **72** (1986), 64-68.
20. K. Gindele, M. Köhl, M. Mast,
    Evaluation of spectral hemispherical reflection measurements
    in the infrared and their application to rough surfaces.
    Proc. of SPIE Symp. on Optical Materials Technology for
    Energy Efficiency and Solar Energy Conversion V,
    SPIE Vol. **653** (1986), 260-266.
21. J. A. Jacquez and H. F. Kuppenheim,
    Theory of integrating sphere. J. Opt. Soc. Am. **45** (1955),
    460-470.

GONIOMETRIC MEASUREMENTS OF REFLECTANCE AT SOLAR WAVELENGTHS AND OF EMITTANCE AT THERMAL WAVELENGTHS

Barrie W Jones, James W Gannon, and Ajit A DeSilva
Physics Department, The Open University, Milton Keynes, MK7 6AA, UK

INTRODUCTION

The modelling of various devices that utilize solar energy is facilitated by knowledge of the angular distribution of
(i)   the solar radiation scattered by various surfaces in the devices
(ii) the thermal radiation emitted by the same surfaces.

The scattered distribution is given (at each wavelength λ) by the bidirectional spectral reflected radiance distribution function (BRDF) ρ'', and the emitted distribution (over all wavelengths) by the directional total emittance ε'. We have measured ρ'' and ε' for a number of materials of relevance to solar energy devices.

We define ρ'' as follows [1]

$$\rho'' (\theta_i, \phi_i, \theta_r, \phi_r; \lambda) = dL_r (\theta_i, \phi_i; \theta_r, \phi_r; \lambda)/dE_i(\theta_i, \phi_i; \lambda) \qquad (1)$$

where the incident (i) and extent (r) angles are shown in Figure 1, where E_i denotes irradiance (Wm^{-2}) tagged with the incident direction (θ_i, ϕ_i), and where L_r denotes radiance ($Wm^{-2} sr^{-1}$) often called intensity. Note that L_r is per unit <u>projected</u> area of surface i.e. $dA \cos\theta_r$ in Figure 1.

We define ε' as follows [2]

$$\varepsilon' (\theta, \phi; T) = e' (\theta, \phi; T)/e_b'(\theta; T) \qquad (2)$$

where the angles are the extent angles in Figure 1, where T is sample absolute temperature (K), where e' is the radiant power emitted per unit area of <u>actual</u> surface, per unit solid angle centred on (θ, ϕ) ($Wm^{-2} sr^{-1}$), and where e_b' is e' for a black body

$$e_b' (\theta; T) = \sigma T^4 \cos\theta/\pi \qquad (3)$$

where σ is Stefan's constant ($Wm^{-2}K^{-4}$). We have no reason to believe that ε' varies with ϕ for any of the materials studied here, so henceforth ϕ is dropped.

Corresponding to ρ'' and ε' are the hemispherically averaged quantities $\rho' (\theta_i, \phi_i; \lambda)$ and $\varepsilon(T)$.

DIRECTIONAL TOTAL EMITTANCE $\varepsilon (\theta, \phi; T)$

The sample is held at $T = (368\pm1)K$ in a chamber with blackened walls that are cooled to ≈77 K to reduce unwanted radiation. The detector "sees" the sample through a small aperture in the chamber, and through further apertures that define the angular resolution as $\approx0.7°$ in θ, the setting accuracy being $\approx \pm0.5°$. The radiation incident on the detector is "chopped" <u>en route</u> and the detector output feeds a phase sensitive system to further attenuate unwanted radiation. The effect of residual unwanted radiation is estimated, and small corrections are applied. The sample is turned to various appropriate values of θ, and $\varepsilon'_S(\theta,T)$ (where "s" denotes sample) is measured with respect to $\varepsilon'_R(15°,T)$ of a reference sample of Nextel-2010 black velvet calibrated by the National Physical Laboratory (UK). Sample and reference sample temperatures are the same to within ± 1.5K. Further details are in DeSilva and Jones [3].

Table 1 lists the materials studied, and Figures 2,4,5 show our measured values of $\varepsilon'_S(\theta,T)$. The standard random error in $\varepsilon'_S(\theta,T)$ is $\simeq \pm 0.0002$ at $\varepsilon'_S(\theta,T)$ $\lesssim 0.2$, rising to $\simeq \pm 0.005$ at $\varepsilon'_S(\theta,T) \gtrsim 0.8$. There is also a systematic error in $\varepsilon'_S(\theta,T)$ that is unlikely to exceed $\simeq \pm 3\%$ of $\varepsilon'_S(\theta,T)$. Table 1 also lists our values of $\varepsilon_S(T)$ obtained by integration of $\varepsilon'_S(\theta,T)$, along with a combined random and systematic error that includes the integration error. Values of $\varepsilon_S(T)$ from other sources are also listed in Table 1, and on the whole there is satisfactory agreement.

Figure 2 shows that $\varepsilon'_S(\theta, T)$ for the glass sample closely follows that for an ideal dielectric with the reasonable value of 2.0 for the refractive index n. Furthermore, Figure 3 shows $\varepsilon_S(T)/\varepsilon'_S(0,T)$ versus $\varepsilon'_S(0,T)$ for all materials studied, and our values for glass are very close to the ideal dielectric curve. Also, our values are not very different from the experimental values of Meinel and Meinel [4].

Figure 4 shows $\varepsilon'_S(\theta,T)$ for the metal samples. The general form is as expected for a metal [2] and the values are also in good agreement with previous work [2]. Moreover, Figure 3 shows that our values for aluminium are similar to those of Meinel and Meinel [4].

Values of $\varepsilon'_S(\theta,T)$ for the two non-selective absorbers (Table 1) are shown in Figure 2. At large θ the values of $\varepsilon'_S(\theta,T)$ are larger than for an ideal dielectric, particularly for Solarcoat-50. This could be because we are "seeing" some emission from the metal substrates (Table 1 and Figure 4). Alternatively, or additionally, scanning electron micrographs and surface roughness measurements show that both materials are rough compared to λ, and so in principle some departure from ideal dielectric behaviour is expected. The data in Figure 3 show that Nextel is closer to the ideal dielectric curve than is Solarcoat-50, and this is consistent with the data in Figure 2.

Values of $\varepsilon'_S(\theta,T)$ for the four selective absorbers (Table 1) are shown in Figure 5. All of them have small values of $\varepsilon'_S(\theta,T)$ at $\simeq 368$K, as we expect, and all of them show a slight increase in $\varepsilon'_S(\theta,T)$ with θ up to $\simeq 70°$, similar to the behaviour of metals: these data indicate that the selective behaviour of these materials owes much to the optical thinness at $\lambda \gtrsim 3\mu$m of the dielectric coat on the metal substrate. At small θ the value of $\varepsilon'_S(\theta,T)$ for Skysorb is that of the stainless steel substrate, whereas for the other three materials the values of $\varepsilon'_S(\theta,T)$ at small θ are significantly greater than the substrate values of about 0.03 for nickel and about 0.02 for copper. This could be due to non-negligible optical thicknesses of the dielectric coatings at $\lambda \gtrsim 3\mu$m (but still optically thin). Alternatively, or additionally, this could be due to the known lack of flatness of the metal substrates in at least two of these materials (Maxorb and Cusorb). The values of $\varepsilon_S(T)/\varepsilon'_S(0,T)$ for Maxorb and Skysorb in Figure 3 are close to the ideal metal curve, whereas the values for Cusorb and Solarcoat-100 are less close.

For further discussion see DeSilva and Jones [3].

BDRF ρ'' $(\theta_i, \phi_i; \theta_r, \phi_r; \lambda)$

The sample is placed in the path of a parallel laser beam, broadened to 4 - 5mm diameter. At λ = 633nm an unpolarized laser and a plane polarized laser are used, the plane of polarization of the latter being rotated with a $\lambda/2$ plate. At λ = 1152mm a plane polarized laser is used, and in the absence of a $\lambda/2$ plate a second orientation of polarization is obtained by rotating the goniometer through 90° in a plane perpendicular to the beam. At both wavelengths the sample is rotated in its own plane, though ϕ_i (Figure 1) is defined with respect to the apparatus. Variations in θ_i are obtained by tilting the goniometer, and the exitent direction (θ_r, ϕ_r) is selected by the location of a large area silicon photodiode detector about 70mm from the sample. A circular aperture on the detector defines the acceptance angle as ≈4.5°. The uncertainties in angular positioning are no greater than about ±0.5°, and the overall (random and systematic) uncertainty in ρ'' is about ±3% of ρ''. The laser beam is "chopped", and the detector output is fed to a phase sensitive detector. For further details see DeSilva and Jones [5].

We have obtained ρ' $(\theta_i, \phi_i; \lambda)$ by means of an integrating sphere, using freshly pressed barium sulphate powder as a reference sample.

Our results using unpolarized lasers on all the selective and non-selective absorbers in Table 1 have been described elsewhere [5]. Here we report new data on Nextel and Maxorb. These data are shown in Figures 6 and 7. In all of these Figures the key shows a view of the sample plane. The square box [□] denotes the sample. Maxorb is grooved, and the direction of the grooves is shown inside the box. The centre line (⊕) denotes the plane of incidence, always ϕ_i = 0 (Figure 1). The unpolarized beam is denoted by the dot (·) and the electrical vector of the plane polarized beams by the double headed arrows (↔ or ↕). If θ_i = 0 then the polarization indicators are in the centre of the box, and if θ_i = 40° then they are below the box. If ϕ_r = 0 or 180° then no further indicators are shown, but if ϕ_r = 270° then a single headed arrow is used (→□).

The Nextel sample consists of 4 spray coats on an aluminium substrate. From the integrating sphere we obtain ρ'(3°, ϕ; 633nm) =(2.6 ± 0.1)%, and ρ'(3°, ϕ; 1152 nm) = (2.6 ± 0.1)%, for all ϕ and all polarizations. The unpolarized data on ρ'' at 633 nm and θ_i = 0 in Figure 6(a) show the range of values of ρ'' as ϕ_r is varied in 10° steps from 180° to 360°, and we conclude that Nextel is close to being axially symmetrical. This is borne out by the polarized data in Figure 6(a) and by further polarized data (not shown) at ϕ_r = 240° and 320°. We can also see from Figure 6(a) that the unpolarized data are the mean of the two appropriate orthogonal polarized data. It is also clear that the orientation of the plane of polarization with respect to a given exitent direction can make a large difference to ρ''. This is also the case at θ_i = 40°, as shown in Figure 6(c) (where the unpolarized data have been omitted for clarity, but are again the mean of the polarized data), and at $\lambda = 1152$ nm, as shown in Figures 6(b) and (d). We finally conclude from Figure 6 that Nextel is not a perfect diffuser, particularly at $\theta_i = 40°$ where it exhibits strong forward scattering. This has also been observed by Hsia and Richmond [6].

For Maxorb $\rho'(3°, \phi; \lambda)$ depends on the sample and on the polarization. A representative value at 633 nm is about 2% for the unpolarized laser. Unpolarized data for ρ'' at 633nm are shown in Figure 7(a) and (c). Note that Maxorb is far from being diffuse, and shows a fairly strong specular peak. This is also the case at 1152nm, as shown in Figure 7(b), which is polarized data. Some of the data in Figure 7 are for <u>crushed</u> Maxorb (see Figure caption). Crushing is performed by applying a uniaxial compression of 7.1×10^5 Nm^{-2} for 60 seconds. This has a modest effect on ρ'' at 633 nm and at 1152 nm. The integrating sphere measurements correspond to $\theta_i \simeq 3°$: at 633 nm and at 1152 nm ρ' rose by about 10% on crushing.

We also subjected Maxorb to thermal cycling. This consisted of placing the Maxorb in an oven at 150°C for 30 minutes, then allowing it to cool at laboratory temperatures, then repeating this cycle five times in all. This made little difference to ρ' and ρ''.

REFERENCES

1   F E Nicodemus, J C Richmond, J J Hsia, I W Ginsberg and T Limperis, Geometrical considerations and nomenclature for reflectance, *Report NBS MN-160*, National Bureau of Standards USA (1977)

2   R Siegel and J R Howell, *Thermal radiation heat transfer*, 2nd ed (McGraw Hill, London) (1981)

3   A A DeSilva and B W Jones, The directional-total emittance at 368K of some metals, solar absorbers and dielectrics, *J Phys D Appl Phys 20*, 1102-1108 (1987)

4   A B Meinel and M P Meinel, *Applied solar energy* (Addison-Wesley, London) (1976)

5   A A DeSilva and B W Jones, Bidirectional spectral reflectance and directional-hemispherical spectral reflectance of six materials used as absorbers of solar energy, *Solar Energy Materials 15*, 391-401 (1987)

6   J J Hsia and J C Richmond, A high resolution laser bidirectional reflectometer with results on several optical coatings, *J Res Nat Bur Standards 80A*, 189-205 (1976)

7   M van der Leij, *Ph D Thesis*, Delft University, The Netherlands (1979)

8   H C Hottel and A F Serafim, *Radiative transfer* (McGraw Hill, London) (1967)

9   A H Musa, *Ph D Thesis*, University of Aston, UK (1980)

Table 1.   Materials investigated

| | | $\varepsilon_s(T)$ | | |
Material	Chemical Layering	This work (368K)	Manufacturers	Others
Dielectrics:				
glass	-	0.84 ± 0.04	-	0.85 (373K)[a]
Metals:				
aluminium	-	0.053 ± 0.003	-	0.09[b]
copper	-	0.023 ± 0.002	-	0.02[b]
brass	-	0.038 ± 0.003	-	0.03[b]
Non-selective absorbers:				
Nextel -2010				
black velvet (3M)	Black paint on aluminium	0.94 ± 0.04	0.93 (422K)	0.96 (373K)[c]
Solarcoat-50 (ZUEL)	Black paint on brass	0.81 ± 0.04	0.93	-
Selective absorbers:				
Maxorb (INCO SS)	Oxides of nickel on nickel	0.100 ± 0.005	0.08-0.11 (373K)	0.14 (373K)[c]
Cusorb (INCO SS)	Oxides of nickel on copper	0.095 ± 0.004	0.08-0.15	-
Skysorb (INCO SS)	Oxides of stainless steel on stainless steel	0.133 ± 0.007	0.10-0.14	0.11 (373K)[c]
Solarcoat-100 (ZUEL)	Black paint on copper	0.184 ± 0.008	0.15-0.25	-

[a] van der Leij (7)

[b] Hottel and Serafim (8)

[c] Musa (9)

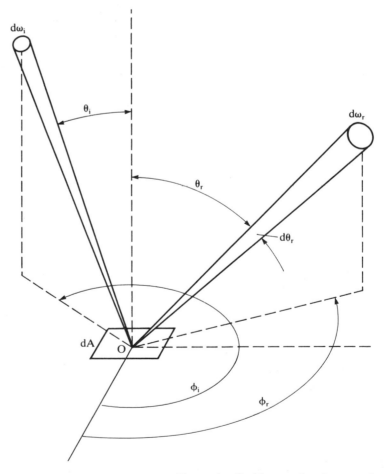

Figure 1   Incident and exitent angles.

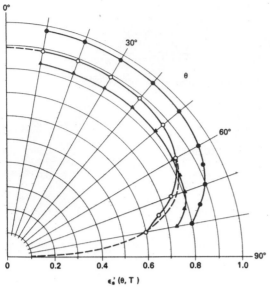

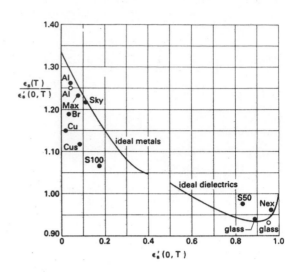

Figure 2 Directional-total emittance for Nextel-2010 black velvet (●), glass (o), and Solarcoat-50 (▲). The broken curve is for an ideal dielectric with n = 2.0.

Figure 3 The ratio $\varepsilon_s(T)/\varepsilon_s'(0,T)$ versus $\varepsilon_s'(0,T)$ for Maxorb (Max.), Skysorb (Sky.), aluminium (Al.), copper (Cu), brass, Cusorb (Cus.), Solarcoat-100 (S100), Solarcoat-50 (S50), and Nextel-2010 black velvet (Nex.). The sources are: this work (●), Meinel and Meinel [4] (o).

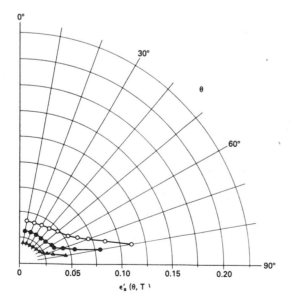

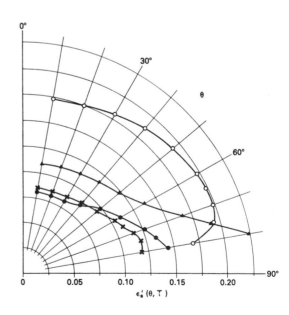

Figure 4 Directional-total emittance for aluminium (o), brass (●), and copper (▲).

Figure 5 Directional-total emittance for Solarcoat-100 (o), Skysorb (▲), Cusorb (x), and Maxorb (●).

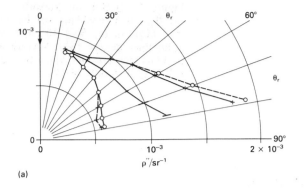

(a)

Figure 6
BRDF for Nextel-2010 black velvet.
(a) $\theta_i = 0$, $\lambda = 633$ nm (b) $\theta_i = 0$,
$\lambda = 1152$ nm (c) $\theta_i = 40°$, $\lambda = 633$ nm
(d) $\theta_i = 40°$, $\lambda = 1152$ nm.
See text for explanation of key.

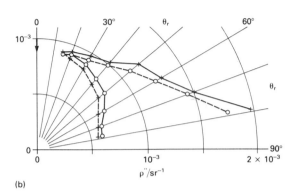

(b)

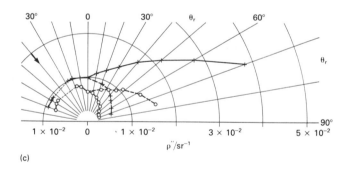

(c)

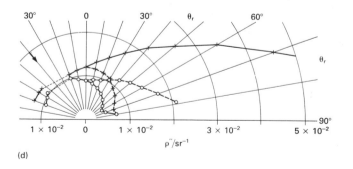

(d)

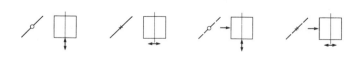

171.

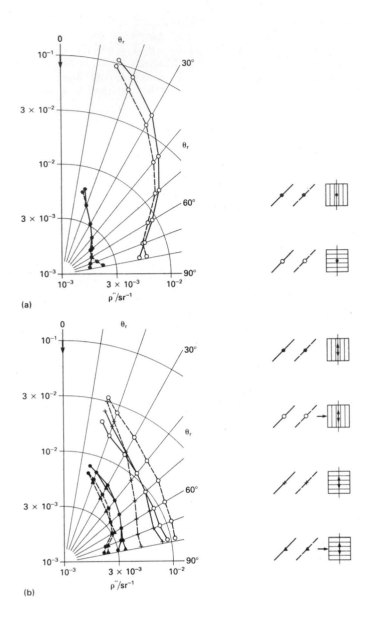

Figure 7 BRDF for Maxorb.
(a) $\theta_i = 0$, $\lambda = 633$ nm (b) $\theta_i = 0$,
$\lambda = 1152$ nm (c) $\theta_i = 40°$, $\lambda = 633$ nm.
Solid lines denote fresh Maxorb and
broken lines crushed Maxorb. See text
for further details of key.

(a)

(b)

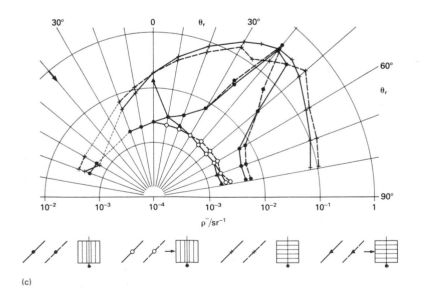

(c)

COMPOSITIONAL DEPTH PROFILING IN SOLAR ENERGY MATERIALS USING MEV ION BEAM ANALYSIS

C Jeynes
Department of Electronic and Electrical Engineering
University of Surrey
Guildford
Surrey
GU2 5XH

DEPTH PROFILING FOR DEPOSITION AND AGEING STUDIES

Modern thin film materials science is making many new materials available to the solar technologist. Because of the low power density of solar radiation, any solar energy material will be produced on a large scale, so cheap deposition processes are required. Many techniques are available, and these are being developed all the time. Having obtained a coating with useful optical properties it is crucial to carry out ageing studies, since the economics of solar installations depend on recovering a relatively high capital cost from a long life with low maintenance costs. Ageing studies will try to describe changes in the thin film due to chemical and thermal stress in terms of, for example: interdiffusion, corrosion, oxidation, hydration, interfacial reactions, permeability and pollutant reactions[1].

Both in depositing a film with desirable properties, and in subsequent ageing tests, it is often important to determine the composition with depth of the film. The ion beam analysis (IBA) techniques[2] can usually provide a quick, accurate, fairly sensitive depth profile with high depth resolution. There are some other standard analytical methods: cross-sectional transmission electron microscopy[3] is very detailed but specimen preparation is always difficult; X-ray photoelectron spectroscopy[4] and Auger electron spectroscopy (AES)[4] can give chemical information and have similar sensitivity to IBA, but because they are sensitive to the true surface they require ultra-high vacuum and are quite slow; secondary ion mass spectrometry (SIMS)[4] is a rapid technique of exceedingly high sensitivity but it requires many standards for quantitative work. The last three techniques depend on erosion of the sample with sputtering ion beams to give a depth profile which may be distorted by the sputtering process. This problem is particularly acute in the presence of interfaces or other sharp concentration gradients. Both SIMS and AES can form quite high resolution images of the surface. IBA requires very little specimen preparation, does not deliberately destroy the sample, can be very quick, requires few standards (it is an absolute technique in principle), and can analyse interfaces accurately.

MEV ION BEAM ANALYSIS TECHNIQUES

A typical IBA instrument (Figure 1) may be used in a variety of different ways. For Rutherford backscattering (RBS)[5], the simplest and most common technique, we would typically used a

1.5 MeV helium beam to irradiate the sample. A few of the
helium particles would suffer elastic collisions with the target
nuclei and be scattered back out of the sample with a large
proportion of their incident energy. These are detected by an
energy sensitive particle detector, and the composition of the
sample with depth determined from the energy spectrum of the
detected particles.

Figure 1
Schematic of an
ion beam analysis
instrument.

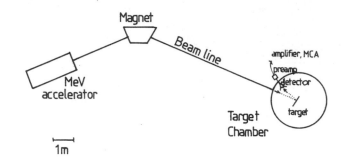

The remainder of the incident beam is implanted into the sample
at a depth much greater than that of analysis, and the energy is
ultimately deposited in the sample as heat. Many inorganic
materials are quite stable under this treatment, but it is
important to remember that although we do not deliberately
destroy the sample during the analysis some samples may be
modified.

The energetic ions can excite K and L shell electrons giving
rise to characteristic X-radiation in the same way that electron
beams do. The advantage of particle induced X-ray emission
(PIXE)[6] is that the X-ray background (mostly bremsstralung) is
orders of magnitude lower than for the electron case since the
proton is so much heavier than the electron. Thus the
sensitivity of PIXE is orders of magnitude better than electron
probe microanalysis. However, the lateral resolution does not
compare since it is difficult and inconvenient to form an ion
probe smaller than a few microns.

Figure 2
Kinematics. Mass M_1
incident with energy E_0
on mass M_2 (stationary)
is scattered through angle
Θ and has final energy E_1.
Recoil of M_2 not shown.

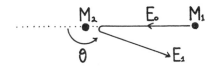

The energies of the particles after a nuclear collision are
determined exclusively by the kinematics, that is: the
conservation of energy and momentum. It is necessary to know
only the energy Q' liberated during the reaction. For an
elastic collision, Q' = 0 and the scattered particle has a final
energy E_1 (see Figure 2):

$$E_1 = kE_0 \tag{1}$$

where $k = [r \cos\Theta + \chi(\Theta)/(1 + r)]^2$ is the kinematical
factor

and $r = M_1/M_2$

$\chi^2(\Theta) = (1 - r^2 \sin^2\Theta)$

and all formulae are in the laboratory frame of reference.
Minimum values of the kinematical factors are obtained for
head-on collisions when $\Theta = \pi$ and $k = (1 - r)^2/(1 + r)^2$. Thus
the best mass resolution is for a high scattering angle.

The energy lost by a high energy beam of particles passing
through a thin film has been measured for many materials (see
Figure 3). Energy loss is quite well understood, and it is
possible to interpolate these measurements to any element[7]. For
compound materials energy losses add linearly. Note that thin
film units of thickness (mg/cm²) are used. This is because thin
film linear thicknesses are difficult to measure absolutely;
besides, density is a function of thickness for thin films. We

Figure 3
$\varepsilon(E)$ for helium
incident on
various target
materials.
(Ziegler 1977)

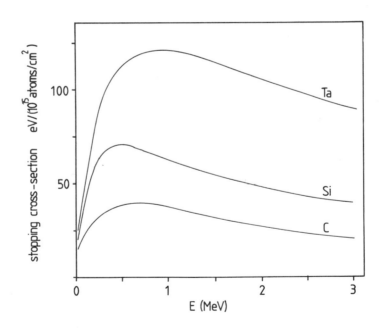

should also emphasise that the natural unit of thickness in IBA
is atoms/cm² (equivalent to mg/cm²) because this is the way
energy loss is measured. For a backscattering spectrum, the
depth scale is given by the change in energy of the particles
scattered from the surface and from the given depth x (see
Figure 4). We define the "energy loss factor" [S]:

$$\Delta = [S]x \tag{2}$$

where $\Delta = kE_0 - E_2$

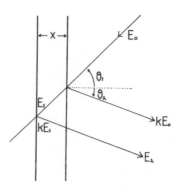

Figure 4
Calculation of the energy loss
factor [S]. A beam energy E_0 is
incident at angle θ_1 onto a thin
film of thickness x. The beam
scattered at angle θ_2 has energies
between kE_0 and E_2. The average
energy of the transmitted beam is E_1

Now because $\varepsilon(E)$ is such a weak function of beam energy E for
MeV beams (see Figure 3), it is clear that [S] is also a weak
function of x. A useful approximation is the "surface energy
approximation" (see Figure 4) where:

$$[S] = k\varepsilon(E_0)/\cos\theta_1 + \varepsilon(kE_0)/\cos\theta_2 \qquad (3)$$

and we can immediately read an approximately linear depth scale
from the energy spectrum. We have included such depth scales in
the examples (Figures 6 to 9); note that the depth scale depends
on the matrix composition, so for strongly varying compositions
the depth scale will be strongly non-linear. These have to be
computed numerically.

In the case of RBS the elastic scattering is due to the Coulomb
fields of the nuclei and the differential scattering cross-
section is analytical:

$$d\sigma/d\Omega = (Z_1 Z_2 e^2/4E)^2 \quad 4(\chi + \cos\theta)^2/\chi\sin^4\theta \qquad (4)$$

Since the probability of scattering given in Equation 4 is
simple and well known it is possible to make absolute
measurements of the atomic concentration: the total yield Y for
any particular energy interval is just given by:

$$Y = Q \, (d\sigma/d\Omega)_{av} \, \Omega \, N \qquad (5)$$

where Q is the number of incident analysis particles, $(d\sigma/d\Omega)_{av}$
is the differential scattering cross-section averaged over the
detector solid angle Ω, and N is the number of particles/unit
area in the slab of material corresponding to that energy
interval. In practice it is usual to avoid measuring $Q\Omega$
directly, instead measuring it with a standard sample. Note the
approximately Z^2 dependence of the yield: RBS is most sensitive
to heavy nuclei.

Where the nuclei have sufficient energy to start to penetrate
the Coulomb barrier, but the interaction is still elastic, the
nuclear structure may modify the scattering probabilities. The
elastic backscattering (EBS) is then non-Rutherford, and EBS

176.

cross-sections may be orders of magnitude higher than RBS ones, and it is often a convenient way to increase sensitivity to the low Z elements.

Elastic recoil detection analysis (ERDA) may be used to detect light elements: an elastic reaction is used but this time low mass forward recoils are detected, with the high mass forward scattered beam stopped differentially by an absorber (Figure 5). This is a convenient way to detect hydrogen which is so important in ageing studies. Note that 2 MeV He ERDA for H is not Rutherford.

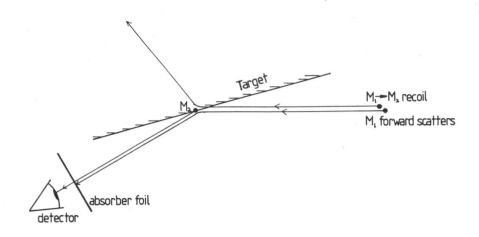

Figure 5   ERDA geometry.
Analysing particles mass M_1 interact in the target. They may scatter equally into or out of the target. If they scatter into the target, the nucleus they scatter from may recoil into the detector. To discriminate forward scattered (heavy) particles from recoil (light) particles an absorber foil is used which is thick enough to stop the heavy particles.

If $Q' \neq 0$ other particles may be produced which may be detected. For example, there are many resonant nuclear reactions producing MeV γ-rays. A depth profile for a particular nucleus would involve stepping the beam energy so that the resonant reaction would occur at different depths in the sample. For example, very good profiles for flourine can be produced in this way with a proton beam of about 400keV.

TWO EXAMPLES

We have recently been pursuing accelerated ageing studies on selectively absorbing black surfaces. In particular we have observed oxidation and reduction of films in various environment, and also some interdiffusion. The aim is to correlate these observations with microstructural and optical properties of the films, and hopefully find recipes for stable films. These films are very rough to improve their optical performance: roughness will clearly degrade the depth resolution of IBA spectra. This will be discussed further elsewhere, we just note here that roughness usually looks like a "diffusion tail" in the composition profile.

Figure 6
1.5MeV He RBS
of black chrome
films electro-
plated onto Ni
plated copper
substrates
annealed in air.
Scattering angle
160°, normal beam
incidence.

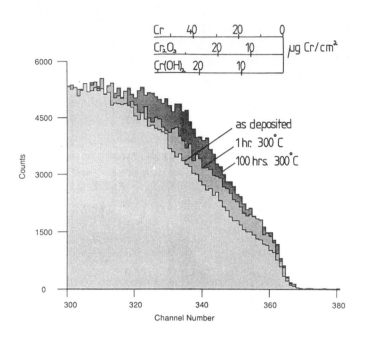

Figure 6 shows RBS spectra from black chrome electro-deposited on a copper substrate with a thin layer of bright nickel. It was suspected that the as-deposited films were hydrated, and the question was, did the ageing process oxidise the film? Clearly, the heat treatment makes the film thinner, not thicker as one would expect if significant oxidation was taking place. We then profiled for hydrogen with ERDA (Figure 7). The spectrum for the 100 hours exposure is representative of a surface contamination peak that one would expect when working in

Figure 7
2MeV He ERDA of the
black chrome films
of Figure 6. Beam
incident at 75°
scattering angle 30°
The absorber was a
6µm polyester film.

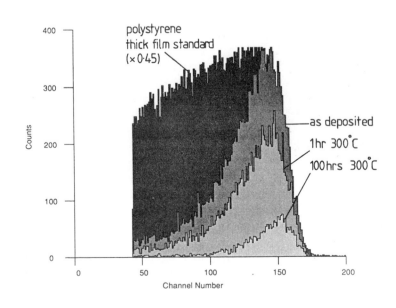

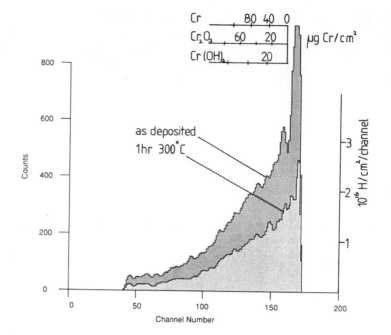

Figure 8
Concentration profiles of H in the black chrome films of Figure 7 (see text).

ordinary high vacuum on very rough surfaces, and is treated as a background. The spectrum is conveniently quantified with a standard specimen of polystyrene (C_8H_8). The cross-section variation with depth is corrected by simply dividing the data by the standard spectrum: the result is shown in Figure 8. The depth scale is obtained exactly as for the backscattering spectra, from a calculation of [S]. The absolute quantification is obtained from [S] for polystyrene together with the observed yield in the polystyrene spectrum. The data above the surface in Figure 8 is an artefact of the division: The surface signal for the polystyrene is at a lower energy than that for the chromium blacks. This is probably due to graphitization at the surface of the polystyrene.

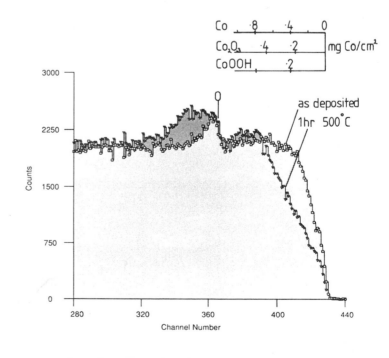

Figure 9
1.65MeV RBS/EBS of blacks formed by chemical oxidation of cobalt metal on nickel plated copper substrates, annealed in air. Scattering angle 160°.

Figure 9 shows a proton backscattering spectrum from black
cobalt oxide films on copper substrates prepared by oxidation of
evaporated cobalt metal films. The cross-section for oxygen
with this analysis beam is about three times Rutherford, which
makes it possible to observe the oxygen profile. RBS is rarely
sufficiently sensitive to oxygen in heavy matrices. It is
obvious that the heat treatment is rapidly oxidising the film.
There is also evidence (not shown) that the copper is diffusing
through the film.

ACKNOWLEDGEMENTS

I would like to thank M G Hutchins for preparing the thin films
selective absorbers described here, and the SERC Central
Facility at Surrey for support of this work.

REFERENCES

1.   See for example: L E Murr, Solar Energy Materials 5 (1981)
      1-19 or C M Lampert, Solar Energy Materials 6 (1982) 1-41.
2.   J W Mayer, E. Rimini (Eds) "Ion Beam Handbook for Materials
      Analysis" Academic Press NY 1977. Many recent examples of
      these techniques can be found in J P Biersack, K Wittmaack
      (eds) "Proceedings of the 7th Int. Conf. on Ion Beam
      Analysis" Nucl. Instruments & Methods B15 (1986).
3.   See for example the proceedings of "Microscopy of
      Semiconductor Materials Oxford April 1987" (IoP Conf.
      Series 87).
4.   See for example current issues of Surface and Interface
      Analysis (Proceedings of "Quantitative Surface Analysis,
      NPL Nov. 1986").
5.   W K Chu, J W Mayer, M A Nicolet "Backscattering
      Spectrometry" Academic Press, NY 1978. Note that with a
      much more sophisticated detector it is possible to do high
      resolution RBS at 100keV with a proton beam: See for
      example J W M Frenken, J F Van der Veen Phys.Rev.Letts. 54
      (1985) 134.
6.   S A E Johansson, T B Johansson Nucl.Instruments & Methods
      137 (1976) 473.
7.   J F Ziegler (Organiser) "The Stopping and Ranges of Ions in
      Matter" (5 Vols) Pergamon 1977.